FLEETING FOOTSTEPS

Revised Edition

FLEETING FOOTSTEPS

Revised Edition

Tracing the Conception of Arithmetic and Algebra in Ancient China

Lam Lay Yong
Ang Tian Se

World Scientific

NEW JERSEY • LONDON • SINGAPORE • SHANGHAI • HONG KONG • TAIPEI • CHENNAI

Published by

World Scientific Publishing Co. Pte. Ltd.

5 Toh Tuck Link, Singapore 596224

USA office: 27 Warren Street, Suite 401-402, Hackensack, NJ 07601

UK office: 57 Shelton Street, Covent Garden, London WC2H 9HE

British Library Cataloguing-in-Publication Data

A catalogue record for this book is available from the British Library.

ISBN-13 978-981-238-696-0
ISBN-10 981-238-696-3

Printed in Singapore

To the scholars known and unknown,
whose footsteps east to west,
have brought about this book.

Contents

Foreword

> "Elementary mathematics is one of the
> most characteristic creations of modern
> thought — characteristic of it by virtue of
> the intimate way in which it correlates
> theory and practice."
>
> — A.N. Whitehead,
> *Essays in Science and Philosophy*
> (1948), p. 132

The title of *Fleeting Footsteps* is drawn from a remark by the historian of mathematics, Florian Cajori. Little more than a century ago, in referring to the various figures of Hindu-Arabic numerals as they evolved over the course of time, he posed a pertinent question: "Need we marvel that, in attempting to harmonize these apparently incongruous facts, scholars for a long time failed to agree on an explanation of the strange metamorphoses of the numerals or the course of their fleeting footsteps as they migrated from land to land?" [Florian Cajori, *A History of Elementary Mathematics*, 1896, p. 14; as quoted on p. xxii of this book].

What Professors Lam and Ang have sought to provide in *Fleeting Footsteps* is an indication as to the source of these footsteps which led to the Hindu-Arabic arithmetic system, a source they identify through a variety of arguments as the ancient Chinese mathematics of the counting rod system of calculation. Not only was this a decimal place-value system, but it evolved algorithmic methods for multiplication and division, root extractions (both square and cube roots), and even the solution of equations on terms very similar to those later adopted by Western mathematicians.

This is not an uncontroversial work, largely because it puts forward in the strongest possible terms, similar claims made as early as the late 19th century by the British sinologist Alexander Wylie, and more recently by Joseph Needham and his collaborator Wang Ling. In volume 3 of the latter's *Science and Civilisation in China*, they note that with a complement of only nine figures, a place-value system existed in China as early as the Shang period (14th century BCE), whereas the earliest surviving evidence of Hindu numerals dates only from the end of the first millennium (CE). As for the transmission of the decimal place-value system beyond China, they also note that "between −250 (BCE) and +1250 (CE) a good deal more [knowledge] came out than went in [to China]," [Needham and Wang, *Science and Civilisation in China*, vol. 3: *Mathematics and the Sciences of the Heavens and the Earth*, 1959, p. 148]. As for the zero, they regard it as "an illustrious invention," one that "seems to have occurred in the very border marches of the two great civilizations, into both of which it quickly spread." Moreover, they go on to suggest that the invention of a symbol for zero was "an Indian garland thrown around the nothingness of the vacant space on the Han counting board," [Needham and Wang 1959, p. 148].

But in *Fleeting Footsteps*, Lam and Ang are interested in much more than the orthography of numerals. What they provide is "an ambitious and multifaceted agenda," as Frank Swetz put it in reviewing the book for *The American Mathematical Monthly* [vol. 161, 1994, p. 921], namely to establish the thesis about which Lam Lay Yong has written over the entire course of her career, but especially so in a series of papers published in the 1980s in such important international journals as the *Archive for History of Exact Sciences, Historia Mathematica, Isis,* and the *Archives internationales d'histoire des sciences*. What she has argued with an increasingly substantial body of evidence is that our modern place-value decimal system and the various methods for multiplication, division, and root extractions in particular, can all be traced back to their earliest forms in the Chinese system of calculation by counting rods using a decimal place-value system.

The text upon which *Fleeting Footsteps* is based is admirably suited to add even more direct evidence to this argument, for in the *Sun Zi Suan Jing* we have the earliest detailed description yet known devoted to the decimal place-value system of counting-rod computation. As I.T. Jakobsen put it, Part I of the book (the introductory material prior to the actual translation of the *Sun Zi Suan Jing*) provides "a very thorough and lucid introduction to the intricacies of the use of the 'counting board' and the rod numeral system... a decimal system where zero was an empty space and the nine numerals were formed from rods or sticks...," [Ivan Taftberg Jakobsen, review of *Fleeting Footsteps*, in *Centaurus*, 36 (1993), p. 366].

But all this, as Alexei Volkov cautioned in reviewing *Fleeting Footsteps* for the *Archives internationales d' histoire des sciences* in 1996, "may provoke a strong reaction from historians of European mathematics." Nevertheless, Volkov emphasizes one of the great strengths of *Fleeting Footsteps* is "the emphasis made by the authors on the great importance of studying Chinese methods of instrumental calculators as well as numerical and algorithmic aspects of Chinese mathematics, which otherwise cannot be understood properly," [Volkov 1996, p. 158].

It is in the final section of Part I of the book that Lam Lay Yong puts forward her thesis in its most compelling and straightforward form: "Like printing, gunpowder and the magnet, three inventions which Francis Bacon said had 'changed the whole face and state of things throughout the world,' the concept of our numeral system should rank as one of China's most significant contributions to human science and civilization," [see p. 185 of this book].

In 2001, in Mexico City, the International Commission on the History of Mathematics awarded its highest honor, the Kenneth O. May medal for outstanding contributions to the history of mathematics, to Lam Lay Yong for her efforts during the course of her career to bring the elements of the history of Chinese mathematics to a larger international audience. This she has done through her numerous publications and translations of such seminal Chinese mathematical works as *A Critical Study of the Yang Hui Suan Fa. A Thirteenth-Century Chinese Mathematical Treatise* (Singapore: Singapore University Press, 1977), and above all in her translation and commentary of the *Sun Zi Suan Jing* that she and Ang Tian Se first published in 1992 in *Fleeting Footsteps*.

Kenneth O. May was the founding editor of the international quarterly journal, *Historia Mathematica*, which began publication in 1974 under the auspices of the International Commission on the History of Mathematics. The prize consists of a bronze medal and a commemorative certificate, and is awarded only once every four years in conjunction with the International Congresses of History of Science. The May medal was first conferred by the International Commission on the History of Mathematics at the XVIIIth International Congress in Hamburg in 1989, when it was jointly awarded to Dirk J. Struik (Cambridge, Massachusetts) and A.P. Youschkevitch (Moscow). In 1993 it was again awarded jointly to Christoph J. Scriba (Hamburg) and Hans Wussing (Leipzig) in Zaragoza, in 1997 it was awarded to René Taton (Paris) in Liège, and in 2001 to Ubiratan D'Ambrosio (São Paulo) and Lam Lay Yong (Singapore) in Mexico City.

When Lam Lay Yong could not be in Mexico City, it was arranged that the medal, along with its accompanying certificate, would be formally awarded to her during a special International Symposium on History of Chinese Mathematics held on August 27, 2002, at the China Science and Technology Museum in Beijing, in connection with the International Congress of Mathematicians. Co-sponsored by the International Commission on the History of Mathematics, the Institute of Mathematics and the Institute for History of Natural Science of the Chinese Academy of Sciences, the day-long symposium included the presentation to Lam Lay Yong of the Kenneth O. May Medal for outstanding contributions to the history of mathematics. Professor Lam followed her acceptance of the May medal with a plenary lecture devoted to the evidence she has amassed in the course of her career for the Chinese origins of the decimal place-value number system as argued in detail in *Fleeting Footsteps*.

As Alfred North Whitehead once observed, "Elementary mathematics is one of the most characteristic creations of modern thought — characteristic of it by virtue of the intimate way in which it correlates theory and practice." What *Fleeting Footsteps* demonstrates is the intricate ways in which practice, the use of counting rods to advance a highly detailed algorithmic series of procedures to facilitate intricate arithmetic computations, including square and cube root extractions, provided a theoretical structure that proved practical and concise, and was easily transmitted, along with the place-value system itself, well beyond China.

Thus it is especially appropriate that *Fleeting Footsteps* is now being reprinted by the World Scientific Publishing Company. This is an important tribute to the contribution that this volume has made to the history of Chinese mathematics. It also testifies to the important contributions Lam Lay Yong has made in the course of her distinguished career, one devoted to bringing to the academic world at large a better and more detailed understanding of the history of Chinese mathematics.

Joseph W. Dauben
July 2003

Visiting Research Fellow, The Needham Research Institute, Cambridge, England;
Professor of History and History of Science,
The Graduate Center, City University of New York, U.S.A.

Preface

Not long after I received the Kenneth O. May Medal at the International Congress of Mathematicians in Beijing in August 2002, Dr K. K. Phua, Chairman and Editor-in-Chief of World Scientific, informed me that he would like to print a second edition of this book.

I decided that an edited version of my plenary lecture given at the above Congress would serve as an appropriate introduction to the book. The aim of the lecture entitled "Ancient Chinese Mathematics and Its Influence on World Mathematics" was to give a lucid understanding of a complex subject, most of which is from this book. The lecture dealt with ancient Chinese arithmetic, the beginnings of algebra and my research findings that the Hindu-Arabic numeral system and its arithmetic originated in China. From the feedback that I received after the lecture, I was glad that the lecture, delivered in the power point format, was easy for the audience to follow and it was well received.

I hope the general reader would agree with me that this edited Beijing lecture serves as a useful introduction to the main work which, understandably, had to be written in a scholarly format.

Except for the Foreword by Professor Joseph Dauben and the lecture text, the rest of the book remain unaltered.

In Part One of the main text, Professor Ang Tian Se wrote most of Sect. 1, the whole of Sect. 8 and provided appropriate insertions for some of the other Sections. In Part Two, he translated the Preface, Qian Baocong's footnotes and gave useful comments on my translation of the text.

I would like to thank Professor Joseph Dauben for the Foreword. I would also like to thank Ms Kim Tan and her team at World Scientific for their efficient help and for giving the book a fresh look.

My husband, Pin Foo, gave numerous comments on the lecture text to enhance its clarity and, as always, his unfailing support.

Lam Lay Yong

Introduction

Today when a person is asked to multiply three thousand five hundred and eight by four hundred and thirty six, he would either use an electronic calculator or calculate on a piece of paper, and produce the solution in a matter of seconds. This task appears simple. The ability to operate a numeral system with instinctive ease is something that we readily take for granted. Some may even presume that the same operation could be performed as quickly with any numeral system, and not just with the Hindu-Arabic numeral system we use today.

In western Europe before the advent of the Hindu-Arabic numeral system, only a limited few would have been able to perform such a multiplication. This may be surprising, but it is only because the Hindu-Arabic system is so ingrained in us that we find it almost impossible to imagine what it would be like to live in a world without it. The primary reason that the peoples of Europe discarded their own numeral systems for the Hindu-Arabic system was that the latter could perform addition, subtraction, multiplication and division far better than any other means of computation that were then known. In the years 1200 to 1600 when the new numerals were gradually adopted, the properties of the system and the methods of performing the four fundamental operations of arithmetic were laboriously learned and practised. A problem such as the above would be taught at university level. According to Karpinski [1925, p. 120], "for several centuries one who could perform long division was considered an expert mathematician".

Thus, when an ancient mathematical treatise expresses in words the multiplication of two numbers and gives the correct answer, this cannot be taken for granted as a simple task. A natural question arises: how was the result derived? How was such a fundamental conceptual breakthrough achieved?

A study of the development of arithmetic in an ancient civilization is best conducted by examining how the four fundamental operations of addition, subtraction, multiplication and division are performed. This, the investigator would soon discover, hinges very much upon the concept of the numeral system employed and the method of computation. These may prove radically different from our Hindu-Arabic system. It is thus important to understand the evolution from within the ancient system, unclouded by preconceived notions derived from the Hindu-Arabic numeral system. This, as would be shown later, is especially important in the case of the Chinese rod numeral system which is not a written numeral system but a method of performing arithmetic with the use of computational rods.

Part One of this book begins with an examination of the concept of the Chinese number system. From this, we show how the early Chinese performed their basic arithmetic operations. We then analyse further arithmetic developments and show how this was extended to the field of algebra.

As far back as the Warring States period (475–221 BC), the Chinese used straight rods or sticks to do their calculation. They formed numerals from the rods, and they did their addition, subtraction, multiplication and division with these rod numerals. The performance of a multiplication problem such as the above with these rods would be commonly known at a very early time not only among mathematicians, but also among officials, astronomers, traders and others. The rods were carried in bundles and, whenever calculation was required, they were brought out and computation was performed on a flat surface such as a table top or a mat. After the results were obtained, they would probably be recorded and the rods would be put away.

Together with an analysis of the Chinese number system, Part One of the book also explains the rod numeral system. This is followed by a step by step explanation of how addition, subtraction, multiplication and division were performed through the rods. These are essential information for anyone wishing to know about traditional mathematics in China.

Almost all ancient Chinese mathematical treatises omitted this basic information most probably because the readers of those times were assumed to be familiar with them. *Sun Zi suanjing* 孫子算經 (The mathematical classic of Sun Zi) written around 400 AD was an exception. The treatise not only gave a description of the rod numerals, it also described the multiplication and division processes and illustrated them with examples.

Our present day arithmetic text describes and is also able to show the procedures of arithmetic operations, because we compute through a written numeral system. On the other hand, the ancient Chinese mathematical text could only describe and give instructions concerning arithmetic operations. The actual computation to obtain such and such a result had to be carried out through the manipulation of rods. The detailed procedures of the operations were understandably not easy to write about and, in most texts, such descriptions were either omitted or were extremely concise. Therefore, when we try to understand a mathematical method of an ancient Chinese script, it is important to understand the text as being intended for use with rod numerals.

One of the best ways to get acquainted with the rod numeral system is through a study of *Sun Zi suanjing*. Part Two of this book provides a translation of the version edited by Qian Baocong 錢寶琮 [1963]. We have chosen this work because, firstly, it was a mathematical primer and, secondly, as we have already mentioned, it contained more descriptions on how to perform arithmetic operations with rod numerals than other works. Furthermore it serves as a central source of reference in our analysis of the evolution of arithmetic and its subsequent extension into algebra. There were other works on this subject which appeared earlier than *Sun Zi suanjing*, the best known and most important being *Jiu zhang suanshu* 九章算術 (Nine chapters on the mathematical art) which appeared about three to four hundred years before the former. However, while *Jiu zhang suanshu* was more advanced, the rudiments of mathematics were not explained, thus making it less suitable as a source of reference for our purposes.

This book has three objectives. The first is to show the development of arithmetic and the initial stages of algebra in China. The second is to conduct a detailed study of *Sun Zi suanjing*. Given the central importance of *Sun Zi suanjing*, the first two tasks are inextricably interwoven.

In Part One of the book, Sect. 1 gives a historical perspective of *Sun Zi suanjing*. After the Chinese number system and the rod numeral system have been explained in Sect. 2, and the fundamental operations of arithmetic in Sect. 3, these are followed by the arithmetic operations on fractions (Sect. 4) and root extractions (Sect. 5). The different tables of measures from *Sun Zi suanjing* are discussed in Sect. 6. The various methods arising from the problems in the book are analysed in Sect. 7. The methods include familiar names such as Rule of Three and Rule of False Position, and others such as the Chinese remainder problem and the solution of a set of simultaneous linear equations. The procedures using rods reveal how the Chinese notated mathematical concepts and, through these notations, were able to solve the problems. A prominent example was how they expressed an equation or a set of equations with rods, and then solved them. Sect. 8 gives some glimpses of the socioeconomic aspects of life in Sun Zi' s time through the subject matter of his problems.

We now come to the third aim of this book: to advance the thesis that the Hindu-Arabic numeral system has its origins in the rod numeral system. The arguments are presented in the last Section (Sect. 9) of Part One since the comprehension of such a thesis requires an extensive knowledge of how the arithmetic and other operations evolved through the rod numeral system.

Lastly a word about the title of the book: *Fleeting Footsteps. Tracing the Conception of Arithmetic and Algebra in Ancient China.* This has been inspired by one of F. Cajori' s [1896, p. 14] remarks concerning the variety of shapes of the Hindu-Arabic numerals: "Need we marvel that, in attempting to harmonize these apparently incongruous facts, scholars for a long time failed to agree on an explanation of the strange metamorphoses of the numerals or the course of their fleeting footsteps as they migrated from land to land?" (See p. 178.) This book is an attempt to trace to the source the "fleeting footsteps" of the concept of the Hindu-Arabic numeral system and, with it, our arithmetic and algebra.

Ancient Chinese Mathematics and Its Influence on World Mathematics[*]

First, let me begin with the title. The term "World Mathematics" will be confined to the arithmetic and the beginnings of algebra that are being taught in our schools today. In this lecture, I shall focus only on the mathematics in ancient and medieval China that had influenced and played important roles in the development of arithmetic and the beginnings of algebra.

The ancient Chinese used bamboo rods to count and, in doing so, they evolved a numeral system. The first nine numerals formed from the rods were

Ⅰ	Ⅱ	Ⅲ	ⅢⅠ	ⅢⅡ	⊤	⊤⊤	⊤⊤⊤	⊤⊤⊤⊤
1	2	3	4	5	6	7	8	9

Note that the horizontal rod represented the quantity 5. For numerals greater than these they had an ingenious device. Numerals in tens, hundreds, and thousands were placed side by side, with adjacent digits rotated, to tell each apart. For example, 1 was represented by a vertical rod, but 10 was represented by a horizontal one, 100 by a vertical one, 1000 by a horizontal

---

[*] Edited lecture delivered at the International Congress of Mathematicians, Beijing 2002.

one and so forth. Zero was represented by a blank space so the numerals 84,167 and 80,167 would be as shown.

$$\text{π} \equiv \text{I} \perp \text{π} \qquad \text{π} \quad \text{I} \perp \text{π}$$

$$8 \ 4 \ 1 \ 6 \ 7 \qquad\qquad 8 \ 0 \ 1 \ 6 \ 7$$

Although most books on the early history of mathematics, especially the recent ones, have mentioned the Chinese rod numerals, they have failed to draw attention to a very important fact that the ancient Chinese had invented a NOTATION. Any number however large could be expressed through this place value notation which only required the knowledge of nine signs. I should add that in the current more sophisticated written form, a tenth sign, in the form of zero, is required. Without notations, a mathematician will be unable to express his ideas effectively. This notation is amongst the earliest that is still in use. Without this notation, the ancient Chinese would not have been able to develop their mathematics and without this notation, our arithmetic would be completely different.

The invention of the rod numeral notation arose from the necessity of doing sums. You will grasp the logic of the formation of the numeral notation when 2 numerals are placed on the board and added, for example 8 + 6.

$$\text{π} \qquad \text{—} \ \text{IIII}$$
$$\text{T}$$

[i]   [ii]

Note that the 2 horizontal rods representing 2 fives are added to become a ten. Note also that the second display would have taken place on the same spot where the first display was as the performer manipulated his rods.

The ancient Chinese used the rod numerals to add, subtract, multiply and divide. For multiplication and division, they needed multiplication tables which were written on bamboo strips. They were similar to ours except that they began with "nine nines are eighty-one" and proceeded downwards.

Through the process of division, the ancient Chinese invented another useful notation, which is still being used today, that is, the notation to express a fraction. For instance, how do we express 4 sixths or 2 thirds? I shall show you briefly the process through which the notation was derived. This is described in *Sun Zi suanjing* 孫子算經 (*The mathematical classic of Sun Zi*) c.400 AD.

Take the example: Divide 100 by 6.

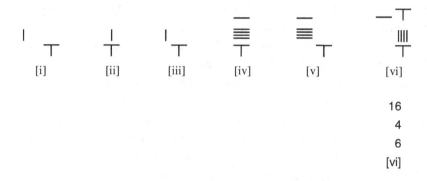

[i]	[ii]	[iii]	[iv]	[v]	[vi]

16

4

6

[vi]

The quotient is placed in the first row, the dividend in the second row. The divisor, in the third row, moves according to the rules.

- [i] shows 1 placed in the hundreds place of the second row and 6 placed in the units place of the third row.
- [ii] shows the 6 in the units place is shifted to the hundreds place. However, 1 above is less than 6 below, division is not possible, so 6 is shifted to the tens place, as shown in [iii].
- 10 divided by 6 gives 1 and remainder 4 as shown in [iv].
- The divisor 6 is next shifted to the units place as shown in [v].
- Six sixes are 36, the quotient 6 is in the units place of the first row and 40 is reduced to 4. The answer is shown in [vi] which is read as 16 and 4 sixths.

I have spread out the division process into six parts. Actually, the changes from the manipulations of the rods would all take place at the first position and they would be performed through recitations. The speed would depend on the skill of the performer. Finally, do note that the notation for a fraction, 4 sixths, is equivalent to our notation.

Try to do problems with fractions without this notation and you would realise the same difficulties that the peoples of other ancient cultures encountered. With the use of this notation, the ancient Chinese developed the subject on fractions to the fullest. *Jiu zhang suanshu* 九章算術 (*Nine chapters on the mathematical art*) c.100 AD (see [Lam 1994])[1] has the following methods: the reduction of a fraction, the addition, subtraction, averaging, multiplication and division of fractions.

With the use of this notation, the methods are similar to the ones that are being taught in our schools although the medium of one is through the use of rods and the other is a written system.

The notation for a fraction in both ancient and modern arithmetics is the outcome of the numeral notation and the division method, and both notations are essential for the full development of arithmetic.

All scholars who know the *Jiu zhang suanshu* would agree that it is a fascinating book. Let me read to you a bird's eyeview of the contents: the common fraction; areas; rule of three; least common multiple; extraction of square and cube roots; volumes; proportion and inverse proportion; relative distance and relative speed; surplus and deficit; rule of false position; the matrix notation; negative numbers; simultaneous linear equations; right-angled triangles.

The description of the methods in this book can be said to be concise. This is understandable since the working of the problems was through the performance of the rods and this was done on any flat surface such as a table or on the floor. I shall refer to this flat surface as "a counting board". The rods were carried in a hexagonal bundle consisting of 271 pieces. Initially, they were about 14 cm long, but by the 6th century they were shortened to half its length. They were not confined to bamboo sticks; they could be made from wood, bone, horn, iron, ivory or jade.

The contents of the *Jiu zhang suanshu* is at such an advanced stage that there are manifestations of the development of arithmetic into what we call "algebra" today. I shall now give you some idea how the ancient Chinese began with the subject.

---

[1] While most of the material in this lecture is taken from the first edition of this book, I have also incorporated sources from four of my other published works. These are referred to in this manner and can be found in the bibliography of this revised edition.

This is a photograph of ivory rods displayed in the renowned History Museum of Shaanxi Province 陝西歷史博物館 in Xi'an.

We all know that the Rule of False Position was a method devised by ancient man to solve a problem, which nowadays is seemingly so simple <u>because</u> it is solved with the right set of mathematical notations. This method was used in ancient Egypt, in medieval Europe and also found in the *Jiu zhang suanshu*. The Chinese called it *ying bu zu* 盈不足 meaning surplus and deficit. We shall take the first problem in Chapter 7 of this book to illustrate how it is solved on the counting board. The problem reads "Now there are a number [of persons] buying goods. If a person pays 8 there is a surplus of 3, if a person pays 7 there is a deficit of 4. Find the number of persons and the cost of the goods. Answer: 7 persons; the goods cost 53".

$$\text{T} \qquad \text{TT}$$
$$\text{IIII} \qquad \text{III}$$

The importance of the method lies in the placement of the rod numerals. The positions of the numerals on the board symbolise what the numerals stand for; namely, the upper positions stand for the proposed values and the lower positions their corresponding surplus or deficit. The positioning of the numerals enables cross-multiplication which is one of the operations in the method.

This placement of rod numerals initiated a mathematical notation on the board to aid the solution of what we call 2 linear equations in 2 unknowns. The way was paved for such solutions and the solutions of other sets of linear equations involving more unknowns. This is found in the next chapter called *fang cheng* 方程.

$$3x + 2y + z = 39$$

$$2x + 3y + z = 34$$

$$x + 2y + 3z = 26$$

I	II	III
II	III	II
III	I	I
$=$ T	$\equiv$ IIII	$\equiv$ IIII

For example, this set of equations using our notations would be represented on the board as shown. Note that each of these equations would have been represented vertically on the board from right to left.

The Chinese solution would involve the manipulation of this matrix which consists of multiplying a column by a suitable number and a series of subtractions between the columns — the end result being a transformed matrix with the remaining numerals forming a sort of triangle as shown below.

			III
		IIIII	II
	$=$ T	I	I
$\perp$ IIII	$=$ IIII	$\equiv$ IIII

This method led to another branch of mathematics — the class of negative numbers. When a column of a matrix was subtracted from another, it was inevitable that negative numbers appeared. A different coloured rod would be used to denote a negative numeral, or alternatively, an additional rod would be placed diagonally across its last non-zero digit. For example, -642 would appear as T≡Ⅺ.

I hope that now you have some idea how arithmetic and the beginnings of algebra were developed on the Chinese counting board. It was inevitable that procedures and algorithms arose which led to a succession of commentators to write them down. There were many commentators on the *Jiu zhang suanshu*, the most well-known being Liu Hui 劉徽 (263 AD).

What began as simple calculations with rods led to the mathematisation of procedures from the counting board onto a written form, and this tradition was carried on continuously for over one thousand two hundred years and resulted in the writing of numerous mathematical treatises.

Because of time constraint, I shall just focus very generally on the development of polynomial equations (see [Lam 1986a]) to give you some notion of how such advanced mathematics could be developed with what, one could say, a mere bundle of rods!

The method of solving a numerical polynomial equation began with the finding of the length of a side of a given square area. A problem in the *Jiu zhang suanshu* on finding one side of a square of area 55,225 showed that the derivation was from geometric considerations but computations were on the counting board, where the positionings of the rod numerals were of great importance. From the diagram shown, they first considered the area in white, then the shaded area and lastly the dotted area (see [Lam 1987a]).

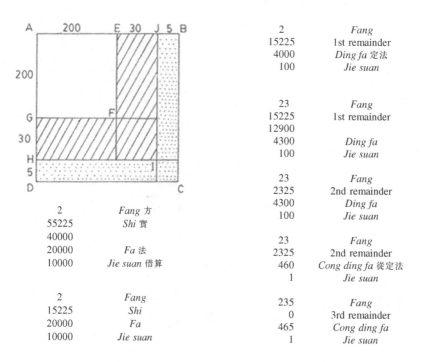

2		Fang
15225		1st remainder
4000		Ding fa 定法
100		Jie suan

23		Fang
15225		1st remainder
12900		
4300		Ding fa
100		Jie suan

23		Fang
2325		2nd remainder
4300		Ding fa
100		Jie suan

23		Fang
2325		2nd remainder
460		Cong ding fa 從定法
1		Jie suan

235		Fang
0		3rd remainder
465		Cong ding fa
1		Jie suan

2		Fang 方
55225		Shi 實
40000		
20000		Fa 法
10000		Jie suan 借算

2		Fang
15225		Shi
20000		Fa
10000		Jie suan

This is to give you a very general idea of the method, the answer as you can see is 235.

From books on the history of mathematics, we know that mathematicians throughout the centuries strove to find solutions for numerical polynomial equations beginning with the quadratic and cubic equations. The Chinese solved them by the correct positioning of rod numerals on the board. The algorithm method that they developed for solving a polynomial equation of any degree in the 13th century is now accepted as similar to that introduced by Ruffini and Horner at the beginning of the 19th century.

In the year 1247, Qin Jiushao 秦九韶 published *Shushu jiu zhang* 數書 九章 (*Mathematical treatise in nine sections*). He showed in detail how to solve an equation of this form: $-x^4 + 763200\,x^2 - 40642560000 = 0$. The display of the rod numerals in the different stages of calculations took up twenty one panels in his book.

I like to show you a diagram which is familiar to all of you.

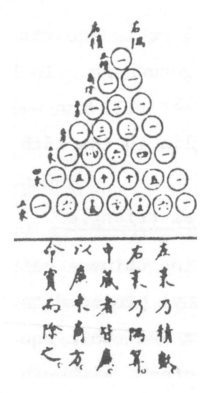

This Chinese representation of what we call "Pascal Triangle" is from *Xiang jie suanfa* 詳解算法 (*A detailed analysis of the methods of computations*) (1261). The author Yang Hui 楊輝 stated that it originated from Jia Xian 賈憲 (c.1100) whose work is no longer extant (see [Lam 1980]).

Let me show you further the wonders that the Chinese mathematicians did with their bundle of rods. Zhu Shijie 朱世傑 in his work *Si yuan yu jian* 四元玉鑒 (*Jade mirror of four unknowns*) (1303) showed how a set of 4 equations in 4 unknowns which are here depicted in our notations was reduced to one equation in one variable.

$$(y-x)^2 + (z-x)^2 + (z-y)^2 + (z-\overline{y-x})^2 + (x+y-z)^2 - 3u + u^2 + u^3 - 2u^4 = 0$$

$$(x+y)^2 - 2xy + x + y + z = u^4 + z^2 - y$$

$$\tfrac{1}{2}(x+y+z) + x + y - z = u^3$$

$$x^2 + y^2 = z^2$$

$$2006u^{14} - 11112u^{13} + 22292u^{12} - 19168u^{11} + 2030u^{10} + 12637u^9$$

$$- 8795u^8 - 8799u^7 + 19112u^6 - 9008u^5 - 384u^4 + 1792u^3 - 640u^2$$

$$- 768u + 1152 = 0 \text{ (Ans. } u = 2).$$

Besides these equations, the Chinese mathematicians also developed other branches of mathematics, which are beyond the scope of this lecture.

Let me now point out some of the salient points with regards to the rod numerals. They were used continuously for a long period of time beginning from the Warring States period (475–221 BC) or earlier. There were two distinct groups of users. The first group, which was obviously a very large one, used them purely for calculations, that is, to perform the four fundamental operations of arithmetic. The second group were the mathematicians which was a much smaller group. The development of mathematics through the rods reached its peak in the 13th and 14th centuries. As for the use of the rods for mere calculations, improvements were made to shorten their manipulations from the 8th century onwards. However, they could not satisfy the demands of an advancing civilisation for quicker calculations. This led to the invention of the abacus. By the 15th century, the abacus was already widely used in China.

In the next century, a number of mathematical texts using abacus calculations as the main means of computation were written; among them, the most influential was Cheng Dawei's 程大位 *Suanfa tongzong* 算法統宗 (*Systematic treatise on arithmetic*) (1592). Many regarded this book as an important work because it used abacus calculations instead of rod numerals. I view it from another angle — that it brought about an end to the development of Chinese mathematics that began with the rod numerals. The use of the abacus to shorten the time of calculation necessitated the rote learning of numerous mathematical methods. As a result, the rigorous step by step reasoning so essential to the development of mathematics was discarded, and inevitably, mathematics declined.

In the second part of this lecture, I would like to present to you my arguments and supporting evidence regarding my thesis that the numeral system that is universally used today, commonly spoken of as the Hindu-Arabic numeral system, has its genesis in the Chinese rod numeral system.

Let me begin by giving you a brief account of what led me into an investigation of this nature. I was impressed by the depth of mathematical knowledge that was accumulated from the treatises beginning with the *Jiu zhang suanshu* up to those of the 13th and 14th centuries. Furthermore, they were similar to the mathematics that were developed through the numeral system we use today. The reason for the decline of Chinese mathematics after the 14th century was because it underwent a change of foundations, from mathematics based on rod numerals with its step by step reasoning, to mathematics based on the abacus with its emphasis on learning by rote method.

A statement that kept coming to my mind was this: Had the Chinese, instead of switching to the abacus, transferred the rod numeral system into a written system, then their numeral system would be identical to our numeral system and their mathematics would have continued to develop to greater heights.

The next step of my research was to investigate the origins of the Hindu-Arabic numeral system. The Arab scholars never claimed that they originated the numeral system. The Arabic texts that described the numerals and through which they were transmitted to Europe, called them Indian or Hindu numerals. They are the earliest existing texts on the subject, and based on this information the Western scholars called the numeral system "Hindu-Arabic".

Despite this name, there has not been any convincing evidence that our numeral system has its origins in India. We have Euclid's *Elements* validating the beginnings of our geometry, *Jiu zhang suanshu* depicting the arithmetic of ancient China, but there is no early Indian text or evidence explaining the origins of the numeral system, the four fundamental operations of arithmetic and the subsequent development of arithmetic.

Scholars who had hypothesized an Indian origin had admitted that they were baffled as to how the system originated. They included T. Dantzig, D.E. Smith & L.C. Karpinski, K. Menninger and F. Cajori. The latter also drew attention that there were three investigators, working independently of one another, who had questioned the assumption of an Indian origin. Furthermore, the Indians themselves did not know how the Hindu-Arabic numeral system originated. B. Datta & A.N. Singh said that the inventor of the system was unknown.

Besides our present numeral system, the Chinese rod numeral system is the only other numeral system that uses the place value notation whose base is 10. In other words, this numeral notation is not found in any other ancient numeral systems including the Indian numeral system.

Let me show you a sample of the ancient numeral systems.

Babylonian cuneiform numerals

Y	YY	YYY	𒐘	𒐚	𒐛	𒐜	𒐜	𒐝	𒐞
1	2	3	4	5	6	7	8	8	9	9

◀	◀Y	◀◀◀	Y◀◀Y	Y◀◀◀
10	11	30	81	100

The Babylonian cuneiform numerals have a place value notation but the base is 60. This system was developed very early, around 3000 to 2000 BC. With such a large base the system was cumbersome to use; a symbol for zero appeared around 3rd century BC but was not used consistently.

Mayan numerals

The Mayan numerals also have a place value notation but the base is 20. As you can see, the 19 signs are quite straightforward and there is a symbol for zero, which looks like the shape of an eye. You can see it at the sign for 20.

Egyptian hieratic numerals

Others, such as the Egyptian hieratic system, do not have the place value notation. They therefore require a system of symbols to represent the numbers 1,2,...,9; 10,20,...,90; 100, 200,...,900; and so on.

Greek alphabet numerals

α	β	γ	δ	ε	F	ζ	η	θ
1	2	3	4	5	6	7	8	9

i	κ	λ	μ	ν	ξ	o	π	Ϙ
10	20	30	40	50	60	70	80	90

ρ	σ	τ	υ	φ	χ	φ	ω	ϡ
100	200	300	400	500	600	700	800	900

The Greek alphabetic system is similar in this respect, and so are the two systems of ancient Indian numerals from the Kharosthi and the Brahmi scripts.

Kharosthi numerals

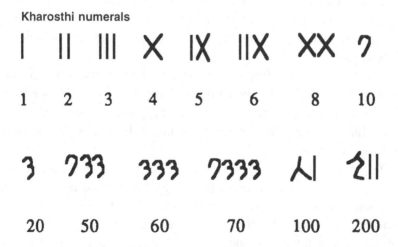

Brahmi numerals

—	=	≡	⅄	⌐	⫝	⁊	⅄	⊋
1	2	3	4	5	6	7	8	9

∝	σ	⌐	×	J	⊣	⁊	Φ	⊕
10	20	30	40	50	60	70	80	90

?	?	?	?	?	?
100	200	500	1,000	4,000	7,000

All these numeral systems require a person to have a fantastic memory when writing large numerals.

Peoples from all over the world adopted the present numeral system, not because they needed a numeral system (as they already had their own), but because they needed the simple place value notation of base ten to learn how to add, subtract, multiply and divide, irrespective of how large the numerals were.

In *Sun Zi suanjing* is found the following: Multiply 708,588 by 531,441 to obtain 376,572,715,308. When this is divided among 354,294 persons, each person gets 1,062,882. Just try to do this problem with, for example, Roman numerals or with numerals from any of the systems that have just been shown to you!

The Chinese invented the decimal place value notation in their numeral system through the use of rods and through this method they developed arithmetic. The arithmetical procedures played a vital role in the advancement of all fields that required calculations. The notations and procedures, which manifest such remarkable clarity and simplicity, place them amongst the greatest achievements of mankind.

Through our present numeral system, we developed arithmetic, and by using the same notation as used by the Chinese for a fraction, arithmetic was further developed. Besides being essential for the advancement of mathematics, we know how necessary arithmetic is for our everyday living. As civilisations progressed from their primitive stages, the numerous advantages manifested by this numeral system were realised by different nations at different times.

In China, during the Tang dynasty (618–907 AD) all civil officials and military officers carried their bundles of rods with them wherever they went, in the same way as we carry our pocket calculators today. Historians generally agree that China, during the Tang dynasty, was the most powerful, most advanced and the best governed country in the world. The splendours of China were a magnet, which drew a continuous stream of foreign visitors to its capital, Chang'an, now Xi'an, which was the largest and most cosmopolitan city in the world.

As the Chinese rod numeral system had been in continuous use for almost two thousand years by Chinese astronomers, mathematicians, scholars, court officials, Buddhist monks, traders and others, the methods of calculation were easily available to foreigners eager to learn new ideas and inventions from the Chinese. These would include foreign traders and other travellers along the busy Silk Road and those resident in China. It is therefore highly likely that some of them adapted this Chinese numeral system into a written system to suit their own requirements.

Between 1200 to 1600, the peoples of Europe began to realise the enormous advantages of the new numeral system. They discarded their own numeral systems in favour of the new one. The methods of performing the four fundamental operations of arithmetic especially multiplication and division were laboriously learned and practised at university level.

The next question you would ask me is: inspite of all these exposures to Chinese influences, could not our universal numeral system have been an independent invention?

I have already mentioned that it was the arithmetic that developed from the decimal place value notation of the numeral system that proved to be of immense use to mankind. The next stage of my research led me to look into Chinese, Indian and Arabic texts which describe the processes of addition, subtraction, multiplication and division for comparison purposes.

Amongst the Chinese texts, *Sun Zi suanjing*, as you will recall was written around 400 AD, is the earliest extant treatise that not only gave a description of the rod numerals, but also described the multiplication and division processes.

The earliest known text on the arithmetic based on our numeral system was written by Muhammad ibn Mūsā al-Khwārizmī around 825 AD. A Latin translation done in the 12th century is in Cambridge University Library. A German translation of this was published by K. Vogel in 1963 and an English translation was done by J.N. Crossley and A.S. Henry in 1990.

When I compared the step by step procedures of using rod numerals for multiplication and division in Sun Zi's book with those procedures described in al-Khwārizmī's work, I was astonished to discover that both sets of procedures were identical.

Let me describe to you briefly the similarities in the multiplication methods. In Sun Zi's book, he began with a general description of how to multiply 2 numbers, no matter how large they were. This was followed by an example: Multiply 81 by 81.

This is my interpretation of Sun Zi's method with rod numerals. As I read out my translation of the method could you please follow it with the rod numerals. The method says: Set up the two positions:[upper and lower] [i]. The positioning of the 2 numerals is explained in the general method. Briefly, the units digit of the multiplier in the third row should be below the first digit from the left of the multiplicand placed in the first row. The translation continues as follows:

- The upper 8 calls the lower 8: eight eights are 64, so put down 6,400 in the middle position [ii].
- The upper 8 calls the lower 1: one eight is 8, so put down 80 in the middle position [iii].
- Shift the lower numeral one place [to the right] and put away the 80 in the upper position [iv].
- The upper 1 calls the lower 8: one eight is 8, so put down 80 in the middle position [v].
- The upper 1 calls the lower 1: one one is 1, so put down 1 in the middle position [vi].
- Remove the numerals in the upper and lower positions leaving 6,561[vii].

Once again please note that although we have seven diagrams here, the person multiplying the numerals would be continuously changing the rods at one place, that is, at the first diagram. Note also that only the multiplier in the third row can be moved from left to right successively.

In al-Khwārizmī's work, there is also a general description of multiplication followed by an example. This is Crossley and Henry's translation of the example, multiply 2,326 by 214.

> An example of this is: When we wished to multiply two thousand three hundred XXVI by CCXIIII, we put two thousand three hundred XXVI by means of Indian symbols into IIII places, and in the first place, that is on the right, there was VI, and in the second two, that is XX, and in the third three, that is three hundred, and in the fourth two, that is two thousand. After this we placed under two thousand IIII, then in the preceding place toward the left one, which is X, then in the third two, which is two hundred, and this is their form:
>
> /   2326\
> \214   /
>
> [Fol. 107v] After this begin from the last place above and multiply it by the last place of the lower number, which is under it. And what

results from the multiplication, write up above. Next write also in the place that comes next by returning toward the right of the lower number. Then do likewise, until you multiply the last place of the upper number by all the places of the lower number. And when you have completed this, transfer the lower number one place toward the right. And the first place of the lower number will be under the place that comes after the number that you have multiplied toward the right. Then put the rest of the places successively; after this also multiply the number itself, under which you put the first place of the lower number, by the last place of the lower number; then by that which comes next, until you have done them all, just as you did in the first place. And whatever is accumulated from the multiplication of each place, write it in the place above it. And when you have done this, transfer also that number, that is your own, by one place and do with it just as you did in the first places. And do not cease from so doing, until you complete all the places. And thus multiply the whole upper number by the whole lower number.

Kurt Vogel translated the passage into German and added this precise interpretation from the long-winded passage.

2326	**428**326	**428**326	**492**226
214	214	214	214

496486	**497764**
214	

It is easier and quicker to show you how multiplication was done from his interpretation. The placement of the multiplier and the multiplicand is the same as the Chinese method. The upper 2 multiplies the lower 2 and the product 4 is placed above the lower 2. In the same way, the upper 2 multiplies the lower 1 and then the lower 4. The resulting product is in darker print. The next step is the shifting of the lower number, 214, one place to the right as shown in the third diagram. The upper 3 then multiplies the lower 2 and the product is added to the number above. In the same way, the upper 3 multiplies the lower 1 and then the lower 4. The same process is then repeated with the shifting of the lower number one place to the right. This goes on till the final product, 497,764, is obtained.

Please note that this was a written system so that every stage had to be written down, whereas the Chinese performed the method at one spot with the rods. You will notice that the Chinese and the Arabic methods are the same, except that the first two rows of the Chinese method are merged into the first row of the Arabic method. This difference is of little significance and it is highly possible that the change was made on the board by the later Chinese.

In the method of division, the Chinese and the Arabic methods are again identical. In the interest of time, I shall very quickly show you my translation of the method in *Sun Zi suanjing* of dividing 6561 by 9 followed by my interpretation of it with rod numerals. The method of division is the same as the simple example I showed you earlier: 100 divided by 6. Similar to that example, the divisor here is placed in the third row and is shifted to the right in stages until the answer, 729, is obtained. There is no remainder here.

Method: First set 6,561 in the middle position to be the *shi* 实 (dividend). Below it, set 9 persons to be the *fa* 法 (divisor) [i]. Put down 700 in the upper position [ii]. The upper 7 calls the lower 9: seven nines are 63, so remove 6,300 from the numeral in the middle position [iii]. Shift the numeral in the lower position one place [to the right] and put down 20 in the upper position [iv]. The upper 2 calls the lower 9: two nines are 18, so remove 180 from the numeral in the middle position [v]. Once again shift the numeral in the lower position one place [to the right], and put down 9 in the upper position [vi]. The upper 9 calls the lower 9: nine nines are 81, so remove 81 from the numeral in the middle position [vii]. There is now no numeral in the middle position. Put away the numeral in the lower position. The result in the upper position is what each person gets [viii].

[i]	[ii]	[iii]	[iv]
[v]	[vi]	[vii]	[viii]

Next, this is part of the description, translated by Crossley & Henry, of how to divide 46468 by 324 from al-Khwārizmī's book, followed by Vogel's summary of it shown at the bottom. Note in Vogel's display that the divisor, 324 in the third row, is being shifted to the right successively for the next digit of the quotient in the first row. Finally, look at its last section which gives the answer as 143 and the fraction 136 over 324. As with the multiplication method, all stages of the division method had to be laboriously written down.

In division on the other hand put the number that you wish to divide in its places; next put the number itself by which you wish to divide under it. And let the last place of the number by which you are dividing be under the last place of the upper number that you are dividing. Moreover, if the number that is the last place of the upper number that you wish to divide is less than that which is the last place of the lower number by which you are dividing, retract the place itself toward the right, until the number of the upper number is greater, i.e., put the last place of the lower number by which you are dividing under the second place that comes after the last place of the upper number. After this consider the first place of the number by which you wish to divide and put in its column above the upper number that you are dividing or beneath it in its column some number which, when you have multiplied it by the last place of the lower number by which you are dividing, will be the same as that number that was in the upper place or close to it, provided it is less than it. When you know it, multiply it by the last place of the lower number and subtract what results to you from the multiplication from that which is above it, from the upper number that is being divided. Once again multiply it by the second place that comes after the last place toward the right, and subtract it from that which is above it and proceed in the division just as you proceeded in the beginning of the book, when you wished to subtract some number from another number; and likewise proceed until you multiply it by all the places of the lower number by which you are dividing. After this move all the places of the lower number, by which you are dividing, one place toward the right....

1	1	14	14	14
46468	14068	14068	2068	1268
324	324	324	324	324

14	143	143	143	143
1108	1108	208	148	136
324	324	324	324	324

I also found the same description of the multiplication and division methods in two other very early Arabic books which have been translated into English. They are: Al-Uqlīdisī's *Kitāb al Fusūl fī al-Hisāb al-Hindī*, 952 and Kūshyār ibn Labbān's *Kitāb fī Usūl Hisāb al-Hind*, c1000.

There are numerous ways of performing multiplication and division through the numerals. In fact, in the Arab countries and Europe, after the Hindu-Arabic numeral system and the four fundamental operations of arithmetic were understood and learned, other methods of multiplication and division began to appear. These were more conducive to a written numeral system. The earliest methods of multiplication and division were suitable for the use of movable rods and naturally not conducive for a written numeral system. That the earliest methods in the Arab countries and those of ancient China are precisely the same is a remarkable phenomenon and cannot, in my view, be attributed to coincidences.

In view of my foregoing arguments, I believe I have adduced sufficient evidence to support my thesis that our numeral system and its arithmetic have its genesis in the ancient Chinese rod numeral system and its arithmetic.

As the overwhelming majority of people do not read ancient Chinese mathematical texts, so, with the collaboration of Professor Ang Tian Se, I translated *Sun Zi suanjing* into English in our book, *Fleeting Footsteps. Tracing the Conception of Arithmetic and Algebra in Ancient China.* The substances mentioned earlier supporting my thesis are described in detail in this book. The book was published in 1992 and to date I have not received any dissenting views on my thesis.

Finally, let me give you a brief history of the use of zero and the other nine symbols of our numeral system. The earliest written numerals had a blank space like the rod numerals when there was no digit of a particular rank. The vacant space was called *kong* 空 in China, *sunya* in India and *sifr* in the Arab countries, and all three words meant empty. It was only later that the empty space of the written system was replaced by a dot or the zero symbol.

As for the nine signs of our numerals, it is generally believed that they can be traced to the first nine Brahmi numerals. As you have already seen, the numeral system does not have a place value notation.

Actually, the history of the shapes of the Hindu-Arabic numerals had been long and tortuous. In fact, there were more than one version of the numerals found. It was the advent of printing that largely formalised the shapes that we are now familiar with.

In conclusion, let me sum up the evidence that our universal numeral system has its genesis in the Chinese rod numeral system.

1. The earliest known methods of multiplication and division were identical with the earliest recorded Chinese methods.
2. Both numeral systems have a place value notation of base ten. They are the only known numeral systems with this notation. The Indian numeral system from which the first nine symbols were derived does not have this notational property.
3. The Arab scholars never claimed they originated the numeral system, while the Western and Indian scholars were unable to provide convincing evidence of its Indian origin.
4. The Chinese used the rod numeral system continuously for almost 2000 years. The system and the accompanying arithmetic so essential for the progress of a nation were easily available to those foreigners, who were able to grasp their significance and adapted them to suit their own needs.

PART ONE

The *Sun Zi Suanjing*
(The Mathematical Classic of Sun Zi)

1.1   The text in perspective

The third century AD saw the vigorous growth and brilliant achievements of mathematics in China. The two mathematical texts of antiquity, namely, *Zhou bi suanjing*[1] 周髀算經 (The mathematical classic of the Zhou gnomon) and *Jiu zhang suanshu* 九章算術 (Nine chapters on the mathematical art),[2] were by then properly annotated by Zhao Junqing 趙君卿 and Liu Hui 劉徽 respectively. The former mathematician highlighted the Pythagorean triplets by supplying geometric-algebraic proofs of their relationships, and the latter elucidated the ancient Chinese mathematical concepts and methods with his own emulations and ramifications (see [Gillon 1977, Swetz & Kao 1977, Ang 1978, Lam & Shen 1984, Wu ed. 1982]). These were followed by *Sun Zi suanjing* 孫子算經 (The mathematical classic of Sun Zi) with the author's preface, which elaborated on the importance and functions of mathematics.

Unlike *Zhou bi suanjing* and *Jiu zhang suanshu* which epitomize the accumulation of mathematical knowledge for a period of time, *Sun Zi*

---

[1] The romanization of Chinese characters in this book is based on the *hanyu pinyin* 漢語拼音 (Chinese Alphabet Phonetic) system.

[2] *Jiu zhang suanshu* has been translated into Russian by Berezkina [1957] and into German by Vogel [1968].

suanjing reflects the effort of one who had benefited from the experience of his predecessors. Unlike Liu Hui and others who made their contributions by enriching and embellishing the existing texts, Sun Zi 孫子 saw the need for a basic text aimed at introducing the mathematical tables and the basic approach to mathematical operations. Although *Sun Zi suanjing* could not be compared with *Jiu zhang suanshu* in terms of mathematical content, yet it is able to stand on its own in terms of historical and mathematical significance. It provides a valuable source for understanding the art of computing with rod numerals, and it has the famous remainder problem of indeterminate analysis (Ch. 3, Prob. 26). The text gained official recognition when it was incorporated as one of the ten mathematical classics commissioned for use in the imperial examinations in 656 AD.

The present version of *Sun Zi suanjing* consists of three chapters and a short preface. In the preface, Sun Zi placed mathematics on the pedestal of knowledge and portrayed it as the regulator of nature and mundane affairs. In his eager attempt to highlight the varied functions of mathematics, Sun Zi even said that mathematics could be used "to locate the positions of the celestial and terrestrial spirits" (p. 190). This tinge of the supernatural function of mathematics may be taken as an indication that Sun Zi might indeed have been a monk or a religious scholar.

1.2   Author and date

The identity of Sun Zi[3] or Master Sun has yet to be confirmed. In the bibliographical chapters of the *Sui shu* 隋書 (Standard history of the Sui dynasty) (656 AD), the *Jiu Tang shu* 舊唐書 (Old standard history of the Tang dynasty) (945 AD) and the *Xin Tang shu* 新唐書 (New standard history of the Tang dynasty) (1061 AD), *Sun Zi suanjing* was listed, but the author's name was not mentioned. This shows that as early as the middle of the 7th century, no one seemed to know who Sun Zi was.

It is obvious that Sun Zi had neither high political position nor influential social standing to merit a place in official history. He appeared to be merely a scholar with some Buddhist inclinations. This is evidenced by a problem

---

[3] Zi was an honorific expression given to a respectable person irrespective of whether his full name was known or unknown such as Kong Zi (Confucius), Lao Zi, etc.

on the length of a sutra (Ch. 3, Prob. 4); it is likely that he had in his possession a copy of the sutra as he was familiar with the number of characters in each chapter. Since the introduction of Buddhism to China from the beginning of the Christian era, it was not uncommon for Buddhist monks to be well versed in mathematics. For instance, in the 5th and 6th centuries, a good number of monks and hermits pursued mathematics in seclusion.[4]

Not only was the authorship of *Sun Zi suanjing* difficult to establish, the dating of the text, too, was equally baffling. Several assumptions were based on internal evidence of the text. The mention of Chang'an 長安 in Ch. 3, Prob. 33 confirmed that the text could not have been written during the late Zhou 周 period in the 3rd century BC as assigned by Ruan Yuan 阮元 (1764–1849) or as put forward by Zhu Yizun 朱彝尊 of the 17th century [Qian ed. 1963, p. 275]. This is because Chang'an was first established as a capital and named as such by Emperor Hui 惠, who reigned in the early years of the Han dynasty. The appearance of the problem concerning a Buddhist sutra, which has been mentioned earlier, also confirmed that *Sun Zi suanjing* could not be a pre-Qin text, since Buddhism was not brought to China until the following Han dynasty.[5]

Since there is a problem on the chess board (Ch. 3, Prob. 5) and another on household taxation (Ch. 3, Prob. 9), Yan Dunjie [1937] made a study of the chess game called *wei qi*[6] 圍棋 and the system of taxation. He found that the chess board of 19 lines with a total of 361 positions on the grid and

---

[4] An example of such a trend is apparent in the biography of Yin Shao 殷紹 in *Wei shu* 魏書 (Standard history of the Wei dynasty), [Ch. 91, p. 1955].

[5] Dai Zhen 戴震 (18th century scholar and mathematician) held this view. Ruan Yuan was aware of Dai Zhen's argument in dating the text to the Han dynasty. However he believed that the problem concerning Chang'an and the one on the Buddhist sutra together with others such as the one on the pregnant woman (Ch. 3, Prob. 36) were probably later additions or deliberate amendments of Sun Zi's original version. It was for this reason that he assigned the date of the text to the late Zhou period.

[6] *Wei qi* literally means "encirclement chess" and another name for it is *go* which is an abbreviation of its Japanese name. The game is still played today; it has become an international game and interest in it is still growing. The game is played on a board which has 19 horizontal lines and 19 vertical lines, forming 361 intersection points. The players use black and white chips or chessmen which are placed at the intersection points. In ancient and medieval China, *wei qi* used to be an intellectual pastime among court officials and scholars.

the system of household taxation had their inception during the middle of the 3rd century AD. Yan inferred from this that *Sun Zi suanjing* could have been a product of the Wei-Jin period (220–420 AD).

In the preface of *Zhang Qiujian suanjing* 張邱建算經 (The mathematical classic of Zhang Qiujian) the author, Zhang Qiujian 張邱建, singled out the method of the problem on "floating of bowls" in *Sun Zi suanjing* (Ch. 3, Prob. 17) for emulation (see [Ang 1969, pp. 120–121]). In the same preface, Zhang also mentioned a problem on "granary" in *Xiahou Yang suanjing* 夏侯陽算經 (The mathematical classic of Xiahou Yang). While the work of Zhang could be ascertained to be written between 468 and 486 AD, that of Xiahou Yang 夏侯陽 remained controversial. In his analysis of the textual content, Qian Baocong [1964, p. 79] commented that apart from the operational rules of arithmetic and the definition on division at the beginning of the text, the rest of *Xiahou Yang suanjing* was probably written by Han Yan 韓延 in the 8th century. The present version does not contain the so-called "granary" problem nor does it carry the commentary of Zhen Luan 甄鸞 (c. 570 AD) as listed in the bibliographical chapters of *Sui shu*, *Jiu Tang shu* and *Xin Tang shu*. When Zhang mentioned the solutions of both Sun Zi's and Xiahou Yang's problems, he seemed to refer to them as contemporaries.

Taking into consideration the text of *Sun Zi suanjing* as a whole, Qian [ed. 1963, p. 275] believed that the present version had been tampered with slightly and the original text was probably written at the turn of the 5th century.

In his discussion on the dating of *Sun Zi suanjing*, Wang Ling [1964, p. 489] said that it could not have been written earlier than 280 AD and later than 473 AD. This was because the taxation method by family units (*hu diao* 户調) in terms of silk floss (*mian* 綿) (Ch. 3, Prob. 9) was established in 280 AD; as for the later date, that was the time when the mensuration scale between *chi* 尺 and *duan* 端 was changed, and Sun Zi followed the old scale (see also [Needham 1959, p. 33]).

1.3 Existing versions

When a department of mathematics was instituted in the National Academy in 656, mathematics was included as a subject in the official examinations. Li Chunfeng 李淳風 was commissioned to edit with annotations a series of

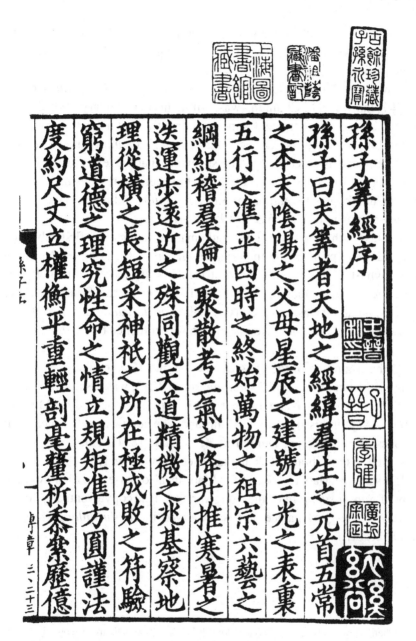

度約尺丈立權衡平重輕剖毫釐析秒忽歷億

窮道德之理究性命之情立規矩准方圓謹法

理從橫之長短采神祇之所在極成敗之符驗

迭運步遠近之殊同觀天道精微之兆基察地

綱紀稽群倫之聚散考二氣之降升推寒暑之

五行之准平四時之終始萬物之祖宗六藝之

之本末陰陽之父母星辰之建號三光之表裏

孫子曰夫箕者天地之經緯羣生之元首五常

孫子箕經序

Fig. 1.1 The first page of the Preface of *Sun Zi suanjing* from the Southern Song edition, now in the Library of Shanghai. At the top of the page, the first seal from the left indicates "Library of Shanghai"; Mao Jin's 毛晉 seal is displayed on the right immediately below the title.

ten mathematical texts for examination purposes. *Sun Zi suanjing* was one of the ten texts known collectively as *Suanjing shi shu* 算經十書 (Ten mathematical classics).

It was noted in the bibliographical chapter of *Xin Tang shu* [Ch. 59, p. 1547] that when Li Chunfeng was writing his commentary, he had in his possession a copy of the text annotated about a century earlier by Zhen Luan 甄鸞. In all subsequent versions of *Sun Zi suanjing*, the front page was almost always preceded by the statement that the text was "officially annotated by Li Chunfeng and others". However, for reasons unknown to us, Li Chunfeng's annotations of *Sun Zi suanjing* did not survive except for one comment (see p. 65 & p. 195, fn. 4).

The text with the other official mathematical classics was first published in wood block print by the Imperial Library in 1084 during the Northern Song dynasty. This was the earliest printed version of the text. It was reprinted by Bao Huanzhi 鮑澣之 in 1213 during the Southern Song dynasty.[7] Later it was also preserved in the *Yongle dadian* 永樂大典 (Great encyclopedia of the Yongle reign), which was compiled between 1403 and 1407.[8]

At the beginning of the Qing dynasty in mid 17th century, a Southern Song printed copy of *Sun Zi suanjing*, dated 1213, was discovered in the family collection of Wang Jie 王杰. This copy was secured by Mao Jin 毛晉 and kept in his Jiguge 汲古閣 library. The copy was subsequently in the possession of Zhang Dunren 張敦仁 (1754–1834). This copy is now preserved in the Library of Shanghai. (See Figs. 1.1 & 1.2.)

It is said that Mao Yi 毛扆, son of Mao Jin, also possessed a handwritten copy of *Sun Zi suanjing* of the Song version. This copy eventually found its way to the Qing palace and is preserved in *Tianlu linlang congshu* 天祿琳琅叢書.

---

[7] See Cheng Dawei's 程大位 *Suanfa tongzong* 算法統宗 (Systematic treatise on arithmetic) (1592) in the section on bibliography.

[8] There are now only three problems of *Sun Zi suanjing* in the existing fragments of the *Yongle dadian*. These are Ch. 2, Probs. 19 & 20 and Ch. 3, Prob. 14 in *Yongle dadian* [Ch. 16344, pp. 10b & 15a; Ch. 16343, p. 17a].

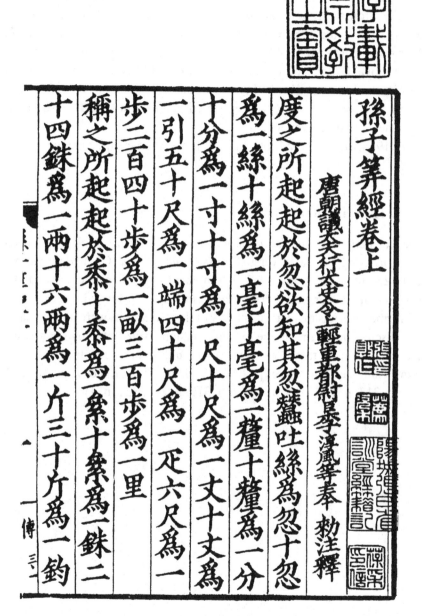

Fig. 1.2 The opening lines of Chapter One of *Sun Zi suanjing*. The first and third seals below the title belongs to Zhang Dunren 張敦仁.

In 1773 when Dai Zhen 戴震 assumed the editing work for the *Siku quanshu* 四庫全書 collection, he extracted and collated the text of *Sun Zi suanjing* from the *Yongle dadian* for inclusion in the encyclopedia compiled between 1773 and 1782 under the patronage of Emperor Qianlong 乾隆. All subsequent versions of *Sun Zi suanjing* such as the *Wu ying dian* 武英殿 block movable-type edition, the *Wei po xie* 微波榭 edition by Kong Jihan 孔繼涵, the *Zhi bu zu zhai* 知不足齋 edition by Bao Tingbo 鮑廷博 and the *Gujin suanxue congshu* 古今算學叢書 edition by Liu Duo 劉鐸 followed Dai Zhen's collated version.

In 1954 Li Yan [1954, pp. 112-125] made some supplementary notes on the text.

1.4   The translation

The translation of *Sun Zi suanjing* in Part Two of this book is based on Qian Baocong's collated version of *Suanjing shi shu* [Qian ed. 1963, pp. 279-322]. Qian [ed. 1963, p. 277] pointed out that when he compared Dai Zhen's collated version with the *Tianlu linlang congshu* edition, he found the former was unsatisfactory as a few of the corrections were unnecessary. The discrepancies among the existing versions were noted by Qian in his annotations on the text; there are 36 notes and we have translated the more significant ones. These are [p. 299, n. 1; p. 303, n. 2, 4, 6 & 7; p. 309, n. 1 & 2; p. 313, n. 1; p. 315, n. 1 & 2; p. 319, n. 2; p. 321, n. 1].

chapter	2

Numbers and Numerals

2.1   Why the need for a detailed study of numbers and numerals?

We first distinguish the use of the two phrases "number words" and "number symbols". A concept of numbers has to be expressed in a spoken language, and when numbers are written in that language, we call them "number words". Thus the number word for "forty three" would be written in the same medium as the word, say, for "table".

There is very often an alternative way of depicting numbers, and this is through the use of notations. When numbers are represented by notations, we call them "number symbols" or "numerals". Numerals, especially large ones, are invented through a concept of numbers. Some examples of number symbols in the ancient world are the Egyptian hieroglyphic numerals, the Indian Kharosthi and Brahmi numerals, the Greek alphabet numerals and the Roman numerals. The obvious examples of number symbols in our society are the Hindu-Arabic numerals.

It is remarkable that peoples of almost all countries throughout the world understand and use the numeral system, 1,2,3,..., which is commonly known as the Hindu-Arabic numeral system. The countries are not unified by a common language so the verbal expressions of numbers may be incomprehensible among the peoples, and the number words differ from place to place; yet in the understanding and application of the numerals, all are tuned to the same wavelength. This phenomenon only occurred during the last few hundred years; prior to this, the various countries had their different systems of numerals analogous to the babel of languages.

The Hindu-Arabic numeral system is not only used for recording numbers, it is an ingenious invention through which numerous computations can be performed. We know that it is the mainstay in the development of our arithmetic. Most significantly it can perform the four basic arithmetic operations of addition, subtraction, multiplication and division far better than any other numeral system. This was the primary force that motivated countries all over the world to adopt the numeral system.

A study of the development of arithmetic in any civilization before the advent of the Hindu-Arabic numeral system should commence with an examination of the number and numeral systems from which the subject evolved. Otherwise we would be using our preconceived notions of arithmetic derived from the Hindu-Arabic numeral system to comprehend and assess the ancient arithmetic. We cannot assume that the ancient concepts of numbers and numerals are identical with the Hindu-Arabic numerals. On the contrary, the latter system possesses special characteristics which make it superior to other numeral systems. The significance of the Hindu-Arabic system is not merely that the numerals are simple to use: what is more important is that the system facilitates computation, and is able to generate numerous mathematical concepts and methods of calculation. Other systems were at best only able to attain some of the concepts and methods.

We shall now discuss the systems of number words and numerals which provided the foundation on which arithmetic developed in ancient China.

2.2   The written numbers

2.2.1 Structure and concept

We begin by analysing the written number system that is found in all Chinese mathematical treatises. Although the number characters are ancient and can be traced to the Qin and Han periods (3rd century BC onwards) (see [Needham 1959, p. 5; Cheng 1983, p. 175; Martzloff 1988, pp. 160–164]), the system is still currently in use.

The structure of the system consists of characters for the first nine numbers and numbers in powers of ten. These are shown below with the romanized script on the right and the corresponding English number words and Hindu-Arabic numerals on the left.

one	1	一	*yi*
two	2	二	*er*
three	3	三	*san*
four	4	四	*si*
five	5	五	*wu*
six	6	六	*liu*
seven	7	七	*qi*
eight	8	八	*ba*
nine	9	九	*jiu*
ten	10	十	*shi*
hundred	10^2	百	*bai*
thousand	10^3	千	*qian*
ten thousand and so on	10^4	萬	*wan*

A number such as 74,385 is written as 七萬四千三百八十五[1] and spoken in the same order, that is, *qi wan si qian san bai ba shi wu*. The literal translation is "seven ten thousand four thousand three hundred eight ten five". A number such as 74,085 is written as 七萬四千八十五 (*qi wan si qian ba shi wu*). The numbers are thus constructed according to the spoken form such that a character representing a number in a power of ten is sandwiched by two characters of the first nine numbers. Occasionally one comes across an exception to this rule for numbers 11 to 19. They may be written as 十一 (*shi yi*), 十二 (*shi er*), ..., 十九 (*shi jiu*) instead of 一十一 (*yi shi yi*), 一十二 (*yi shi er*), ..., 一十九 (*yi shi jiu*). This is because the spoken form has been abbreviated.

Let us analyse the concept of this number system. A number is made up of quantities expressed in terms of numbers in powers of ten. For instance, in our first example, the number is made up of seven ten thousands, four

---

[1] In a Chinese text, the characters are written in columns from top to bottom in the direction from right to left. In our text, the words are written in rows from left to right in the direction from top to bottom. For this reason, we have written a number in Chinese characters linearly from left to right instead of vertically from the top downwards.

thousands, three hundreds, eight tens and five. It is spoken in this order; in the written form, the characters also fall into the same order.

The characters denoting the numbers were evolved from a series of alterations and modifications in writing through a long period of time. They could be traced to the oracle bone numerals of the Shang dynasty (14th to 11th century BC). The tables shown in Figs. 2.1 & 2.2 provide some idea of the change in their shapes [Cheng 1983, pp. 174–175]. Though the shapes of the symbols for the numbers underwent changes, the concept of the number system remained basically unchanged (see [Qian 1964, pp. 5–7, Cheng 1983, p. 178]).

一	二	三	四	五	六	七	八	九	十	百	千	萬

Fig. 2.1 Numerals from the oracle-bone inscriptions of the late Shang period.

一	二	三	四	五	六	七	八	九	十	百	千	萬

Fig. 2.2 Numerals from the stone inscriptions of late Zhou to Han period.

2.2.2 Number word and number symbol are identical

The Chinese language was, and still is, written in ideograms. This means that every concept, idea or thing is represented by a sign or symbol. Because of this, the Chinese number words have a very special feature; they are also number symbols. The number word "seventy four thousand three hundred and eighty five" and the numeral "74,385" are two distinctly different ways of representing the same concept in our context. In Chinese, both form s are denoted by 七萬四千三百八十五. According to Menninger [1969, p. 54], "This accordance of spoken number words and written numerals is unique, the only known instance in world history".

Although our number word (in English) and the corresponding Hindu-Arabic numeral represent the same idea, they were derived from concepts of numbers which originated in different places and at different times. The compactness and convenience of a single representation for both number word and numeral in the unique Chinese case may be better appreciated, when we look into the instances in history where a number word and a distinctly different written numeral were evolved from the same place of origin. The number word of such cases was lengthy, and the corresponding numeral, though concise, did not strictly follow the verbal sequence.[2]

2.2.3 Numerical ranks are indicated

There is another important feature in the Chinese written number system which is also due to its unique structure. The characters for numbers in powers of ten such as 十 *shi*, 百 *bai*, 千 *qian* and 萬 *wan* do not only stand for 10 or ten, 100 or hundred, 1,000 or thousand and 10,000 or ten thousand respectively; they can also mean numerical ranks, which are, in this case, tens, hundreds, thousands and ten thousands respectively.

---

[2] For instance, the Roman numerals did not follow the sequence of spoken words as illustrated below in the number 4,879 [Menninger 1969, p. 53].

four thousand	eight hundred	seventy	nine
quattuor milia	*octingenti*	*septuaginta*	*novem*
(I) (I) (I) (I)	D C C C	L X X	V I I I I

For other examples on the difference between number words and numerals, see [Menninger 1969].

If we denote these ranks for short by T, H, Th and TTh respectively, then the Chinese numeral for 74,385 is composed of 7TTh 4Th 3H 8T 5, and its meaning is 7 ten thousands, 4 thousands, 3 hundreds, 8 tens and 5 units. The correspondence with the characters are as follows:

七	萬	四	千	三	百	八	十	五
7	TTh	4	Th	3	H	8	T	5

A number written in Chinese does not only have the twin features of being both a number word and a number symbol, it also has a third feature of being able to manifest in quantities of different numerical ranks. The quantity of each rank is indicated by a digit from the first nine numbers, which are 一二三四五六七八九. The character immediately following the digit describes the rank. For example, 二百 is 2 hundreds and 二千 is 2 thousands. When a digit is not followed by such a character, then it is in units so that 二 is 2 or 2 units.

In each number, the quantities of the different numerical ranks are arranged in descending order of ranks beginning with the highest.

If a number has no quantity of a certain rank, then that rank is omitted. For example, 70,085 is composed of 7 ten thousands, 8 tens and 5 units and is written as follows:

七	萬	八	十	五
7	TTh	8	T	5

2.2.4 Large numbers

The structure of the number system is such that an extension of the numbers to larger ones requires an extension only in the names for numbers in powers of ten. When the latter names are known, the name of any large number can be formulated. The names of the next three numbers in powers of ten after 10^4 are as follows:

10^5	十萬	*shi wan*
10^6	百萬	*bai wan*
10^7	千萬	*qian wan*

The literal meanings of the characters for these names are obvious, "ten ten thousand" for *shi wan* 十萬, "hundred ten thousand" for *bai wan* 百萬, and "thousand ten thousand" for *qian wan* 千萬. As the characters 十 *shi*, 百 *bai*, 千 *qian* and 萬 *wan* are familiar, these names do not require any effort to remember.

A number such as 83,529,064 is written as 八千三百五十二萬九千六十四 and its transliteration is *ba qian san bai wu shi er wan jiu qian liu shi si*. Note that a single 萬 *wan* placed after 二 *er* is sufficient to describe the ranks of the four digits preceding it.

The names for 10, 10^2, up to 10^7 as listed above were commonly known, and they were generally consistent throughout the ages. The next number, 10^8, is 萬萬 *wan wan* meaning "ten thousand ten thousand" and is also known as 億 *yi*. The first chapter of *Sun Zi suanjing* gives the nomenclature for numbers whose powers of ten are multiples of 8. They are as follows (p. 193):

10^8	萬萬	*wan wan* is called 億 *yi*,
10^{16}	萬萬億	*wan wan yi* is called 兆 *zhao*,
10^{24}	萬萬兆	*wan wan zhao* is called 京 *jing*,
10^{32}	萬萬京	*wan wan jing* is called 陔 *gai*,
10^{40}	萬萬陔	*wan wan gai* is called 秭 *zi*,
10^{48}	萬萬秭	*wan wan zi* is called 壤 *rang*,
10^{56}	萬萬壤	*wan wan rang* is called 溝 *gou*,
10^{64}	萬萬溝	*wan wan gou* is called 澗 *jian*,
10^{72}	萬萬澗	*wan wan jian* is called 正 *zheng*,
10^{80}	萬萬正	*wan wan zheng* is called 載 *zai*.

The numbers in powers of ten lying between any one of the above numbers and the next one have names in an orderly fashion. For example, since 10^8 is called *yi* 億, 10^9 to 10^{15} are shi yi 十億, *bai yi* 百億, *qian yi* 千億, *wan yi* 萬億, *shi wan yi* 十萬億, *bai wan yi* 百萬億, *qian wan yi* 千萬億 respectively. In other words, the characters for 10 to 10^7 are prefixed to 億 *yi*. *Yi* translated into English is "hundred million".[3] The next number, 10^{16}, which is 萬萬億 *wan wan yi*, has another name — 兆 *zhao*. The names

---

[3] The sequence of number names in Chinese is somewhat similar to the names for numbers in powers of ten after a million in the English language. These names are: ten million, hundred million, thousand million, ten thousand million, hundred thousand million, and finally thousand thousand million which is called billion. The cycle of names begins again after "billion". The Chinese has a single name *wan* for "ten thousand", which is absent in

of 10^{17} to 10^{23} are similar to those of 10^9 to 10^{15} with *yi* replaced by *zhao*. The procedure for names of large numbers in powers of ten continues in this structural pattern.

It is thus seen that the system of numbers expressed to whatever size does not require an extensive vocabulary; in fact, the range of characters employed is most economical. When the numbers are below 10^8, it is only necessary to know thirteen characters, namely, the names for 1 to 9, 10, 100, 1,000 and 10,000. When the numbers exceed 10^8, the additional knowledge of the nomenclature for numbers whose powers of ten are multiples of 8 is required.

In describing the above nomenclature for numbers whose powers of ten are multiples of 8 as "the common model of large numbers" (*fan da shu zhi fa* 凡大數之法) (p. 193), it is possible that Sun Zi considered the nomenclature for the above set of large numbers to be commonly accepted. There were variations in the use of the same nomenclature for other sets of large numbers. In fact immediately before this table, Sun Zi used the same names for a different set of numbers. After he had established that the standard measure of one *hu* 斛 had a capacity for 60,000,000 *su* 粟 (grains of millet), he went on to build a series of large numbers in powers of ten. The terminology of this series is as follows (see p. 192):

10^8	億	*yi*
10^9	兆	*zhao*
10^{10}	京	*jing*
10^{11}	陔	*gai*
10^{12}	秭	*zi*
10^{13}	壤	*rang*
10^{14}	溝	*gou*
10^{15}	澗	*jian*
10^{16}	正	*zheng*
10^{17}	載	*zai*

---

the English vocabulary. In the recycle of names, the highest numerical rank in English usage is "thousand", and in Chinese usage is *wan* (ten thousand). This accounts for the English practice of marking off digits in a numeral by three places from the right, while the Chinese number concept would lead to marking off digits by four places. We illustrate this difference in the following example:

English practice: 516,563,659
Chinese practice: 5,1656,3659

Since *Sun Zi suanjing* has the same nomenclature for two different sets of numbers, Qian [ed. 1963, p. 275] had argued that this indicated the book was amended at different periods. While the present text had probably gone through some amendments (see Sect. 1.2), it should be pointed out that in this particular context, there is no inconsistency in Sun Zi presenting two versions of the nomenclature for large numbers. In the early stage, it was quite common for names of large numbers not to have fixed meanings. Even in our era, the British "billion" is different from the American "billion". Moreover the same nomenclature for different sets of large numbers was also recorded in other books. For example, the *Shu shu ji yi* 數術記遺 (Memoir on some traditions of mathematical art), purported to have been written by Xu Yue 徐岳 of the Han dynasty,[4] stated three classes of large numbers called "upper", "middle" and "lower", which used the same nomenclature [Qian ed. 1963, p. 540]. This may be summarised in the following table:[5]

	wan	yi	zhao	jing	gai	zi	rang	gou	jian	zheng	zai
Upper	10^4	10^8	10^{16}	10^{32}	10^{64}	10^{128}	10^{256}	10^{512}	10^{1024}	10^{2048}	10^{4096}
Middle	10^4	10^8	10^{16}	10^{24}	10^{32}	10^{40}	10^{48}	10^{56}	10^{64}	10^{72}	10^{80}
Lower	10^4	10^5	10^6	10^7	10^8	10^9	10^{10}	10^{11}	10^{12}	10^{13}	10^{14}

From this table we see that the group of large numbers which Sun Zi termed as "the common model" corresponds to Xu Yue's "middle" class of numbers.[6] An examination of Sun Zi's use of large numbers in the text indicates that he also followed the nomenclature of this class of number; the largest number is 三千七百六十五億七千二百七十一萬五千三百八

---

[4] Qian [ed. 1963, p. 531] was of the opinion that much of the text was written by Zhen Luan 甄鸞 in the 6th century AD.

[5] Although Xu Yue mentioned the whole set of names from *yi*, *zhao* to *zai*, he only gave the values of each category up to *jing*. The rest of the values shown in the table are inserted according to his principle.

[6] This same class of numbers was quoted by Zhu Shijie 朱世傑 in his *Suanxue qimeng* 算學啓蒙 (Introduction to mathematical studies) (1299). The names were extended to six more numbers, the last number, 10^{128}, was called *wu liang shu* 無量數, which meant "a non-countable number" (see [Lam 1979, p. 8]).

(376,572,715,308) (p. 200). The nomenclature of this set of numbers seems to have been accepted in other mathematical texts as well; *Jiu zhang suanshu* [Qian ed. 1963, p. 155] records the number 一萬六千四百四十八億六千六百四十三萬七千五百 (1,644,866,437,500).

2.2.5 Summary

The Chinese written number system has the unique feature of performing two roles: as a number word system and a numeral system. Furthermore the numbers are built in gradations in units, tens, hundreds, thousands, and so on; the ideographic characters representing numbers in powers of ten also represent numerical ranks. A number can thus be regarded as consisting of various quantities in different numerical ranks, whereby each quantity is expressed by a digit from the first nine numbers followed by the rank. Under this rule of writing numbers, an extension to larger numbers requires an extension only in the names for numbers in powers of ten. The names for such numbers are based on a structure which recycles the names *shi, bai, qian, wan, shi wan, bai wan* and *qian wan* (which stand for 10, 100, 1,000, 10,000, 100,000, 1,000,000 and 10,000,000 respectively), so that new names are only necessary for numbers whose powers of ten are multiples of 8.

The fact that the system has been in use for well over two thousand years, and is still in use, speaks for its usefulness and lasting qualities.

2.2.6 On the translation of numbers in *Sun Zi suanjing*

All numbers in *Sun Zi suanjing* are expressed in the written number system which has been described above. Since number word and numeral are identical, we have a choice of translating a number in the form of a number word or a numeral, and this depends on its context. Most of the number characters in *Sun Zi suanjing* are translated into Hindu-Arabic numerals, though occasionally some are translated into number words. In places where they mean numerical ranks, they are translated accordingly.

We would like to remind our readers to bear in mind the concept of the Chinese number system wherever there is a translation or discussion of numbers from the Chinese texts.

2.3   The rod numerals

2.3.1   Description

The ancient Chinese used the written numbers (or numerals) for recording only and not for computation. When they had to calculate, this was performed through the use of another type of numerals which were formed from straight rods. The rods were carried in bundles and whenever calculation was needed, they were brought out and computation was performed on a flat surface such as a table top, a mat or the floor. Historians have sometimes referred to these rods as "counting rods" and the surface, a "counting board" [Needham 1959, pp. 8–10; Ang 1977; Mei 1983].

The first nine numerals formed from rods were as follows:

I	II	III	IIII	IIIII	T	TT	TTT	TTTT
1	2	3	4	5	6	7	8	9

The rods for each of the above numerals were placed vertically except in numerals 6 to 9, where the single horizontal rod represented the quantity five.

The representation of numerals greater than 9 was ingeniously devised. The value of such a numeral depended on the choice of digits from the first nine numerals and the positions occupied by them on the board. We have seen earlier that the written numerals were built on a gradation of ranks in units, tens, hundreds, thousands, ten thousands, and so forth. The rod numerals were built on the same concept. The difference between the two types of numerals was that while the ranks of the digits in a written numeral were indicated by characters, the ranks of the digits in a rod numeral were indicated by the positions of the digits on a board. A rod numeral such as 34 would have 3 placed to the left of 4 so that 3 occupies the tens' place and 4 the units' place. However since the counting board was any flat surface without marked columns, the representation of 34 by rods in the form IIIIII would obviously be ambiguous. To avoid this kind of ambiguous representation, the Chinese invented another ingenious device. A digit in the tens' place had its rods rotated through ninety degrees, so that vertical rods were turned into horizontal ones and horizontal rods became vertical. Thus 34 in rod numeral notation appeared like this: ≡ IIII. For larger numerals, the same principle of rotating the rods was applied to those digits occupying alternate places to the left of the tens' place.

The nine rod numerals shown above occupied positions whose place values were units, hundreds, ten thousands, and so on; the rotated numerals shown below occupied positions whose place values were tens, thousands, hundred thousands, etc.

一 = 三 ≣ ≣ ⊥ ⊥ ⊥ ⊥

The digits of each numeral were arranged in a horizontal line from left to right in descending order of numerical ranks beginning with that of the highest rank. For instance, the notation of 84,167 appeared as ⊪ ≣ l ⊥ ⊤.

2.3.2 Historical background

According to *Qian Han shu* 前漢書 (Standard history of the Western Han dynasty) [Ch. 21A, p. 956], the rods were cylindrical bamboo sticks, 0.1 *cun* 寸 in diameter and 6 *cun* long. (One *cun* is a Chinese inch approximately equal to 2.31 cm.) By the 6th century AD, the rods were square in cross-section and had become shorter. This is recorded in *Sui shu* 隋書 (Standard history of the Sui dynasty) [Ch. 16, p. 387], which gives the rods as 0.2 *cun* wide and 3 *cun* long. The shorter sticks facilitated easier manipulations and their square cross-sections prevented them from rolling. The *Qian Han shu* also states that the rods were carried in a hexagonal bundle consisting of 271 pieces with 9 rods on each edge. A cross-section of such a bundle is shown in Fig. 2.3.[7]

Fig. 2.3

---

[7] Cheng Chin-te [1925, p. 494] points out that this is an early example of a figurate number.

In the Tang dynasty (618–907), it was quite common for administrators, engineers and military officers to carry their bundles of rods wherever they went (see [Needham 1959, p. 72; Mei 1983, p. 58]). There is no doubt that the earliest material used for making counting rods was the bamboo stick. Owing to its perishable nature, it is difficult to determine precisely the date of invention. Nevertheless it can be safely assumed that the invention would not be later than the 5th century BC because the use of counting rods as an aid in computation during the Warring States period (475–221 BC) was already extremely common.[8]   Subsequent mention of the rods was more frequent. (See [Du 1991, pp. 159–161].) Not only were the shape and size of the rods improved with the passage of time, the material with which the rods were made was not confined to bamboo sticks alone. The rods could be made from wood, bone, horn, iron, ivory or jade.

The more common and lasting ones seem to be those made from animal bones. This is evident in a bundle of counting rods, all made of animal bones, unearthed in a Western Han tomb at Qianyang 千陽 district in Shanxi 陝西 province in 1971 (see Fig. 2.4). The rods, cylindrical in shape, were kept in a silk bag tied around the deceased person. The longest rod is 13.8 cm, while the shortest is 12.6 cm. The majority are approximately 13.5 cm long [Li Di 1984, pp. 58–59]. In 1975, a bundle of bamboo rods were unearthed from a Han tomb at Fenghuangshan 鳳凰山 in Jiangling 江陵, Hubei 湖北 province. These rods are a little longer than the ones of Shanxi province. Three years later, some earthenware of the Warring States period were found at Dengfeng 登封 district in Henan 河南 province. The earthenware have rod numeral signs drawn on them [Li & Du 1987, p. 8]. (See Fig. 2.5.)

Counting rods made of ivory or jade were more for status than for practical use. For example, in the 3rd century AD, Wang Rong 王戎, a minister of state and the patron of watermill engineers, was said to have spent his time, day and night, reckoning his income with his ivory counting rods [*Taiping yu lan*, Ch. 750, p. 2a]. This subsequently led to the proverbial expression, *ya chou ji* 牙籌計 "to reckon with ivory rods", as an allusion to wealth.

---

[8] For example, Lao Zi 老子 in *Daode jing* 道德經 (Canon of the *dao* and its virtues) [Ch. 27, p. 10] said, "Good mathematicians do not use counting rods".

Fig. 2.4   Counting rods unearthed at Qianyang 千陽.

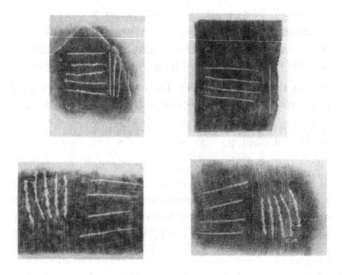

Fig. 2.5   Rod numeral signs on earthenware excavated at Dengfeng 登封.

Sun Zi suanjing is the earliest extant mathematical treatise which gives a description of the rod numerals and the procedures for multiplication and division. These historic accounts provide an insight not only into the formation of rod numerals, but also into how the ancient Chinese performed the fundamental operations of arithmetic. The following description of rod numerals is very often quoted (p. 193).

In the common method of computation [with rods]⁹ (*fan suan zhi fa* 凡算之法), one must first know the positions (*wei* 位) [of the rod numerals]. The units are vertical and the tens horizontal, the hundreds stand and the thousands prostrate; thousands and tens look alike and so do ten thousands and hundreds.

We draw attention to two of the characters in this passage. One of these is *suan* 算 which refers specifically to calculation with rods and the other is *wei* 位 which refers to positions on the counting board. Both are key words to computation with rods, and these terms occur frequently in mathematical texts.

*Xiahou Yang suanjing*¹⁰ gives the same description of the rod numerals and adds the following passage [Qian ed. 1963, p. 558]:

> For 6 and above, [the quantity] five is on top. 6 is not an accumulation
> of rods and 5 is not a single rod.¹¹

This means that for numerals 6 to 9, the single rod on top represents the quantity five; 6 is not represented by ||||| nor ≡, and 5 on its own is not represented by a single rod.

The counting rods were used for addition, subtraction, multiplication and division. They were indispensible whenever calculation was needed, and were in use for a long period of time till the latter half of the Ming dynasty (1368–1644). The invention of the four basic operations of arithmetic through rod numerals provided the foundation for the development of arithmetic. Subsequently the greater part of traditional Chinese mathematics was fostered through the rod system.

---

⁹ In the translation of the text, square brackets [] are used to indicate editorial additions by way of explanation, amplification, or adaptation to the grammar of the English language.

¹⁰ The existing version of this 5th century book was probably reconstructed during the reign of Dai Zong (763–779) in the Tang dynasty (see [Qian ed. 1963, p. 551] and Sect. 1.2).

¹¹ A sentence with the same meaning as the last sentence of this passage is erroneously tucked into a description on multiplication in *Sun Zi suanjing* (p. 194 fn. 3). This once again indicates that the text has been tampered with.

During the Ming dynasty, the demand for quicker means of computation led to the invention of the abacus. The gradual replacement of the rods by the abacus led to the demise of the rod system. However the abacus had neither the potential to foster the growth of mathematics nor the capacity to allow for the conceptual retention of what had already been developed in mathematics through the rod numeral system. Consequently traditional mathematics stagnated and subsequently declined.

2.3.3 The intrinsic properties

We mentioned earlier that a difference between the written numerals and the rod numerals was that the numerical ranks of the digits in the written numerals were indicated by characters while those in the rod numerals were indicated by positions. We now tabulate below the different representations of the number 84,167; first, as a Chinese written numeral followed by its conceptual meaning (see p. 38 for the notation), next, as a rod numeral and, lastly, as a Hindu-Arabic numeral.

Written numeral	八萬	四千	一百	六十	七
Conceptual meaning	8TTh	4Th	1H	6T	7
Rod numeral	𝍤	𝍢	I	⊥	𝍤
Hindu-Arabic numeral	8	4	1	6	7

In both the rod numeral and the Hindu-Arabic numeral, the numerical ranks of the digits are encapsulated in their positions.

The parallelism between the Chinese written numeral and the rod numeral as exhibited in the above example holds for all other numbers. It also holds for numbers which have certain ranks omitted such as in the following examples:

Written numeral	三百七	三千七	三百
Conceptual meaning	3 H 7	3 Th 7	3 H
Rod numeral	III 𝍤	☰ 𝍤	III
Hindu-Arabic numeral	3 0 7	3 0 0 7	3 0 0

A number such as 3,007 has no digits whose ranks are tens and hundreds. When such a number is transcribed into a numeral which uses place values, the places representing these ranks have no digits and are therefore vacant. This gives rise to the concept of what we call zero. This concept is only peculiar to a numeral system which uses place values. The Chinese described the vacant places of a rod numeral as *kong* 空, which meant empty. The original form of the Hindu-Arabic numerals did not have the zero symbol to represent the concept and, like the rod numerals, had a blank space in its stead. This space was called *sunya* in India and *sifr* or *as-sifr* in Islam, and these words meant empty. (See [Karpinski 1925, p. 41; Menninger 1969, pp. 400–401; Dantzig 1930, pp. 29–30].) The Babylonian numeral system, which used place values with sixty as base, also had a blank space to represent this concept of zero. The vigesimal place value system of the Mayan numerals had a symbol which looked like an eye or a closed fist to denote zero. (See [Wilder 1968, p. 51].)

A numeral system, which does not have the place value feature, does not require this concept of zero. Examples of such systems are the Egyptian hieroglyphic and hieratic numerals, the Greek Ionic or alphabet numerals, the Indian Kharosthi and Brahmi numerals (see pp. 174–176), and the Chinese written numerals. For instance, in the Greek alphabet numeral system, $\psi\gamma$ stood for 703, where $\psi = 700$ and $\gamma = 3$. The positions of ψ and γ and the spacing between them did not affect the value of the numeral. (See [Karpinski 1925, p. 13].)

In a place value notation, an absent rank in a numeral is naturally indicated by an empty space in the corresponding rank position because there is no digit of that rank in the numeral. However difficulties may arise when such numerals have to be deciphered at a later time; the number of vacant spaces of a numeral, or where the spaces are, may be unclear. This would then result in an uncertainty in the value of the recorded numeral. Let us consider, as an illustration, the original version of the Hindu-Arabic numerals where there was a vacant space instead of the "0" notation: the numeral, "7 3", could mean 703, or 7,003, and so forth, and "5" could mean 5, or 50, or 500, etc.

A similar situation among rod numerals is slightly better since the digits of a numeral in alternate places would have had their rods rotated. For instance, 703 and 7,003 would not look the same (see above). There is an essential difference between the Hindu-Arabic numerals and the rod numerals: the former are written on a piece of paper but the latter were movable chips. The rod numerals laid on a flat surface during computation were continuously subjected to change, and, in the short span of time, the computer would remember what numbers the rod numerals on the board stood for. The results of the computation would then be recorded in written numerals and these have no ambiguity.

The rod numeral system employed a place value notation with ten as base. It thus required only nine signs and the concept of zero. The Hindu-Arabic numeral system was built on the same concept. The ancient Babylonian numeral system employed a place value notation, but the base was sixty. This required fifty nine signs and the concept of zero. The numerals of the Mayans also used place values but with a vigesimal base. The simplicity and effectiveness of the rod numerals lay in the fact that the place value system needed only nine signs; moreover these signs were easy to remember as they were constructed in a logical manner with each requiring at most five rods.

A numeral with a place value notation is composed of digits occupying certain positions relative to one another; the positions signify the numerical ranks of the digits. In the case of the digits of a rod numeral, they were placed in a horizontal line from left to right in descending order of their numerical ranks. They followed the sequence of the spoken word. Furthermore the rod numerals could be placed anywhere on a flat surface analogous to our writing Hindu-Arabic numerals anywhere on a sheet of paper.

2.3.4 Summary

The difference between the concepts of the written number system and the rod system was that the notion of numerical ranks was depicted by characters in the first system and by positions in the second system. The rod numeral system used a place value notation with ten as base. It required only nine signs which represented the numbers 1 to 9. As each sign was formed from at most five rods, they were simple to understand and easy to remember.

The digits of a numeral were arranged horizontally from left to right in descending order of ranks beginning with the highest. This order of ranks followed the same order of the rhetoric number. The spoken words for ranks were thus transcribed into the positions of the digits on the counting board. The digits in adjacent positions were differentiated from each other by a ninety degree rotation of the rods.

An empty place in a numeral implied that the numeral had no digit of the rank indicated by the position. This blank space standing for what we call zero is a natural consequence of a place value notation.

The rod numeral system was essentially a computing mechanism. The results obtained in the stages of computation were recorded in written numerals which were different from the rod numerals.

The Fundamental Operations of Arithmetic

3.1   Were the operations simple?

Today the skill of performing the four fundamental operations of arithmetic, namely, addition, subtraction, multiplication and division is taught at a very early age in kindergartens and primary schools. This skill is a necessity of everyday living, so it is continuously practised and applied; as a result, it is regarded as simple. The operations are performed through the Hindu-Arabic numerals, and the efficacy of the operations is due to the use of this numeral system.

Why did the peoples in medieval Islam and Europe replace their original numeral systems by the Hindu-Arabic system? Because they needed a numeral system through which they could not only perform the four basic arithmetic operations, but could also perform them with relative ease. If we choose any one of the discarded numeral systems to perform either multiplication or division, we shall soon realise the complexity of the process especially where large numbers are involved. Most ancient numeral systems were built on an intricate structure of requiring a system of notations to represent the following numbers: 1, 2,...,9; 10, 20,...,90; 100, 200,...,900; and so on. Consequently more notations had to be remembered as numbers grew larger. This type of numeral system was found in Egypt, Greece, India and Italy. (See Sect. 9.4 and also [Flegg ed. 1989, pp. 76–101].)

As for those numeral systems with a place value notation, it is often argued that it is easier to compute with a system of base ten than with one of another base, such as sixty, which was used by the Babylonians. However even for the Hindu-Arabic numerals which have ten as base, the initial learning and mastery of how to operate with them were by no means considered simple. In 13th and 14th century Europe, the new arithmetic founded on this numeral system was taught at university level, and an elementary multiplication or division problem needed the expertise of a trained specialist [Karpinski 1925, pp. 54–55, 120].

3.2   Manipulating rod digits

The ancient Chinese used the rods described in Sect. 2.3.1 to perform the four basic arithmetic operations. That they were familiar with these operations by the Warring States period can be seen from a passage of *Fa jing* 法經 (A juristic classic), which was compiled by Li Kui 李悝 who was a minister at the court of the State of Wei (424–387 BC). The passage reads as follows [Li Yan & Du Shiran, 1963, p. 13]:

> A farmer with a family of five cultivates 100 *mu* 畝 of land. Each year one *mu* produces one and a half *dan* 石 of millet, so that the total produce is 150 *dan*. After deducting one tenth of this, which is 15 *dan*, for taxation, there remains 135 *dan*. Each person consumes one and a half *dan* per month, so 5 persons consume 90 *dan* in a year. There is 45 *dan* left. Each *dan* is worth 30 *qian* 錢 so the total worth is 1,350 *qian*. Subtracting 300 *qian* for ancestral sacrifices, leaves a remainder of 1,050. Each person needs 300 *qian* for clothing, so the cost for five persons per year is 1,500. There is hence a deficit of 450.

The numbers were written in the Chinese numerals discussed in Sect. 2.2.1. Before we show how the ancient Chinese performed the four fundamental operations of arithmetic with rod numerals, we shall first discuss the elementary operations of adding and subtracting a pair of rod digits.

The nine digits were notated by rods which were either vertical or horizontal (see Sect. 2.3.1). The addition and subtraction of the digits involved the manipulation of rods based on rules governing vertical and horizontal rods. The vertical rod stood for one kind of unit and the horizontal rod another kind. In the set of digits representing units, hundreds, ten thousands, etc., a vertical rod stood for one unit and a horizontal rod stood for five units as shown below:

$$| \quad || \quad ||| \quad |||| \quad ||||| \quad \top \quad \overline{\top} \quad \overline{\overline{\top}} \quad \overline{\overline{\overline{\top}}}$$

and vice versa for the set of digits representing tens, thousands, hundred thousands, etc.:

$$- \quad = \quad \equiv \quad \equiv \quad \equiv \quad \perp \quad \underline{\perp} \quad \underline{\underline{\perp}} \quad \underline{\underline{\underline{\perp}}}$$

We now examine how the rods were manipulated for the following cases: (a) 8 + 6, (b) 8 − 6, (c) 9 + 2, (d) 7 − 4.

(a) When $\top$ (6) is added to $\overline{\overline{\overline{\top}}}$ (8), the two horizontal rods in the units' place are combined to give one horizontal rod in the tens' place, while all the vertical rods are group together to show four vertical rods in the units' place.

$$\begin{matrix} \overline{\overline{\overline{\top}}} \\ \top \end{matrix} \quad - \quad |||| $$

(b) When $\top$ (6) is subtracted from $\overline{\overline{\overline{\top}}}$ (8), the two horizontal rods of equal amount are removed, leaving three vertical rods as minuend and one vertical rod as subtrahend. Taking one from three leaves two vertical rods as remainder.

$$\begin{matrix} \overline{\overline{\overline{\top}}} \\ \top \end{matrix} \quad \begin{matrix} ||| \\ | \end{matrix} \quad || $$

(c) When $||$ (2) is added to $\overline{\overline{\overline{\top}}}$ (9), there is a total of six vertical rods between the two numerals. Five of them and the horizontal rod of $\overline{\overline{\overline{\top}}}$ are grouped together to yield one horizontal rod in the tens' place leaving a single vertical rod in the units' place.

$$\begin{matrix} \overline{\overline{\overline{\top}}} \\ || \end{matrix} \quad - \quad | $$

(d) When IIII (4) is subtracted from 丅 (7), it is not possible to subtract the four vertical rods from the two vertical rods of 丅 . So subtract the four vertical rods from the horizontal rod of 丅 to leave one vertical rod; this is then added to the two remaining vertical rods of 丅 to show three vertical rods.

丅   III

IIII

The procedure for adding and subtracting two rod digits is based on the simple rules of operating with horizontal and vertical rod representing fives and units respectively, or units and fives respectively, depending on the place values of the digits. The method is a mechanical process easily understood through observation and manipulation.

3.3 Multiplication

3.3.1 The method in *Sun Zi suanjing*

Since there is no explicit account in existing ancient texts on how addition and subtraction are performed with rod numerals, we shall first discuss the methods of multiplication and division, and then infer from them the methods of addition and subtraction. The earliest and most extensive discussion on how to multiply and divide with rod numerals is found in *Sun Zi suanjing*.

The following is the general description of multiplication (pp. 193 – 194). We have fragmented it in order to clarify the general meaning of each part through the illustration of examples.

> In the common method of multiplication (*fan cheng zhi fa* 凡乘之法),
> set up two positions, the upper and lower positions facing each other.
> If there are tens in the upper position then the correspondence is
> with the tens, [i.e., the units of the lower numeral are below the tens
> of the upper numeral].

Let us take the example, 79 multiplied by 63. The rod numerals on the board would be displayed as shown below, with 79 in the upper position, 63 in the lower position, and the "3" of 63 directly below the "7" of 79.

$$\perp \; \overline{\text{III}}$$

$$\llcorner \; \text{III}$$

The text goes on to say:

> If there are hundreds in the upper position then the correspondence
> is with the hundreds, [i.e., the units of the lower numeral are below
> the hundreds of the upper numeral].

We illustrate this by the example, 279 multiplied by 63. The "3" of 63 is
now directly below the "2" of 279 as shown below.

$$\text{II} \; \perp \; \overline{\text{III}}$$

$$\llcorner \; \text{III}$$

The rest of the text is as follows:

> If there are thousands in the upper position then the correspondence
> is with the thousands, [i.e., the units of the lower numeral are below
> the thousands of the upper numeral] [i].[1]

> The upper commands the lower, [i.e., the digit of the upper numeral
> which is above the units of the lower numeral is multiplied by each
> digit of the lower numeral,] and the result is displayed in the middle
> position. When [multiplication with a digit of the lower numeral
> results in a product that] calls out tens, pass the tens over [to the
> left]; the remainder, [i.e., the units,] stays put [in the same column as
> the digit of the lower numeral] [ii & iii].

> Next remove [that digit of] the upper position which has been
> multiplied, and shift one place [to the right] (*tui zhi* 退之) the numeral
> of the lower position which has been the multiplier [iv].

> [In this manner] the numerals of the upper and lower positions are
> mutually multiplied till the process is completed [v to x].

---

[1] The lower case Roman numerals are our additions. They indicate the correspondence of
the description of the operation with the illustration in terms of rod numerals in the example.

We illustrate the complete multiplication process by using the example, 7,239 multiplied by 23.

$$\perp \; \| \equiv \text{π}$$
$$\equiv \text{|||}$$
[i]

$$\perp \| \equiv \text{π}$$
$$- \text{||||}$$
$$\equiv \text{|||}$$
[ii]

$$\perp \| \equiv \text{π}$$
$$- \top -$$
$$\equiv \text{|||}$$
[iii]

$$\| \equiv \text{π}$$
$$- \top -$$
$$\equiv \text{|||}$$
[iv]

$$\| \equiv \text{π}$$
$$- \top \equiv \top$$
$$\equiv \text{|||}$$
[v]

$$\equiv \text{π}$$
$$- \top \equiv \top$$
$$\equiv \text{|||}$$
[vi]

$$\equiv \text{π}$$
$$- \top \perp \| \equiv$$
$$\equiv \text{|||}$$
[vii]

$$\text{π}$$
$$- \top \perp \| \equiv$$
$$\equiv \text{|||}$$
[viii]

$$\text{π}$$
$$- \top \perp \text{|||} \equiv \top$$
$$\equiv \text{|||}$$
[ix]

$$- \top \perp \text{||||} \equiv \top$$
[x]

We now translate the above into Hindu-Arabic numerals.

Upper position:	7239	7239	7239	239
Middle position:		14	161	161
Lower position:	23	23	23	23
	[i]	[ii]	[iii]	[iv]

239	39	39	9
1656	1656	16629	16629
23	23	23	23
[v]	[vi]	[vii]	[viii]

9	
166497	166497
23	
[ix]	[x]

Besides giving the general description of how multiplication is performed, the book also illustrates the method in the first problem of Chapter 1. This is stated as follows:

> Nine nines are 81, find the amount when this is multiplied by itself. Answer: 6,561.
>
> Method: Set up the two positions: [upper and lower] [i]. The upper 8 calls the lower 8: eight eights are 64, so put down 6,400 in the middle position [ii]. The upper 8 calls the lower 1: one eight is 8, so put down 80 in the middle position [iii]. Shift the lower numeral one place [to the right] and put away the 80 in the upper position [iv]. The upper 1 calls the lower 8: one eight is 8, so put down 80 in the middle position [v]. The upper 1 calls the lower 1: one one is 1, so put down 1 in the middle position [vi]. Remove the numerals in the upper and lower positions leaving 6,561 in the middle position [vii].

We now use rod numerals to display the performance of the computation according to the description.

In terms of Hindu-Arabic numerals, the above stages are as follows:

81	81	81	1
	64	648	648
81	81	81	81
[i]	[ii]	[iii]	[iv]

1	1	
656	6561	6561
81	81	
[v]	[vi]	[vii]

We now sum up the description of the multiplication method.

1. In the multiplication of two numerals, one numeral is placed in the first row and called the upper numeral, while the other is in the third row and called the lower numeral.
2. The lower numeral is so placed that its units are aligned with the first digit from the left of the upper numeral.
3. This first digit of the upper numeral is multiplied by each digit of the lower numeral commencing from the left.
4. As each digit of the lower numeral multiplies the upper digit, the product is placed in the middle row such that the units are directly above it and the tens are to the left of the units. Each product is added to the sum of the previous products.
5. After all the digits of the lower numeral have multiplied the digit of the upper numeral, the latter is removed and the lower numeral is shifted one place to the right.
6. The procedure of multiplication continues as before with each successive first digit of what is left of the upper numeral and the lower numeral. The operation goes on till all the digits of the upper numeral are removed. The lower numeral is then removed leaving the result.
7. Although the upper numeral has its digits removed one at a time, the place values of its digits are fixed on the counting board. The place values of the digits of the product in the middle row are aligned with the fixed place values of the digits of this numeral. The place values of the digits of the lower numeral do not correspond to stationary values on the board as the numeral has to be shifted to the right each time a digit of the upper numeral is removed.

3.3.2 The Chinese method and the earliest method in Islam

The earliest method of multiplication using Hindu-Arabic numerals as described by medieval arithmeticians in Islam followed the same step by step procedure as the method stated in *Sun Zi suanjing*. There is a slight difference in the format of the two methods: the Chinese method has three rows and the method described in Arabic texts has two — the first two rows of the former are merged into the first row of the latter. For instance, the Arabic method of multiplying 81 by 81 is as follows:

81	6481	6481	6561	6561
81	81	81	81	

and the Chinese method is as follows (p. 59):

81	1	1		
	648	648	6561	6561
81	81	81	81	

The Chinese also practised the two row multiplication method; we have evidence of this in *Cheng chu tong bian suan bao* 乘除通變算寶 (Precious reckoner for variations of multiplication and division) by the 13th century mathematician Yang Hui 楊輝. (See [Lam 1977, pp. 23–24].)

The Arabic method is described and illustrated in numerous examples of al-Uqlīdisī' s work, *Kitāb al-Fusūl fī al-Hisāb al-Hindī* (952/3 AD). The following is one of the examples [Saidan 1978, pp. 50–51].

If we want to multiply 836 by 79, we insert that as follows: 836.
                                                                      79

Now we multiply eight by seven and insert the units of the outcome above seven and the tens after it. We multiply eight by nine, insert two in place of eight and add seventy, which is seven, to the six.

Then we shift setting nine under three, and seven under two. We multiply three by seven and three by nine.

Next we shift the nine setting it under six, and the seven under seven. We multiply six by seven and six by nine.

In all that we insert the units of the outcome of each multiplication above it and the tens above the tens next to it; whenever a place yields 10, we set it as one thereafter. We talk as we multiply.

The outcome of that is 66044.

The same way of multiplying two numbers is also described in the Latin manuscript of Cambridge University Library Ms. Ii.vi.5 [fol. 107r–108r]. This script is generally accepted as a translation of an Arabic manuscript based on the arithmetic text written by Muhammad ibn Mūsā al-Khwārizmī around 825. As in *Sun Zi suanjing*, a general description of multiplication is first given followed by an illustration of a particular example. We quote below an English translation of the example, which is 2,326 multiplied by 214 [Crossley & Henry 1990, pp. 116–117] .

An example of this is: when we wished to multiply two thousand three hundred XXVI by CCXIIII, we put two thousand three hundred XXVI by means of Indian symbols into IIII places, and in the first place, that is on the right, there was VI, and in the second two, that is XX, and in the third three, that is three hundred, and in the fourth two, that is two thousand. After this we placed under two thousand IIII, then in the preceding place toward the left one, which is X, then in the third two, which is two hundred, and this is their form:

$$/ \quad 2326\backslash$$
$$\backslash 214 \quad /$$

[Fol. 107v] After this begin from the last place above and multiply it by the last place of the lower number, which is under it. And what results from the multiplication, write up above. Next write also in the place that comes next by returning toward the right of the lower number. Then do likewise, until you multiply the last place of the upper number by all the places of the lower number. And when you have completed this, transfer the lower number one place toward the right. And the first place of the lower number will be under the place that comes after the number that you have multiplied toward the right. Then put the rest of the places successively; after this also multiply the number itself, under which you put the first place of the lower number, by the last place of the lower number; then by that which comes next, until you have done them all, just as you did in the first place. And whatever is accumulated from the multiplication of each place, write it in the place above it. And when you have done this, transfer also that number, that is your own, by one place and do with it just as you did in the first places. And do not cease from so doing, until you complete all the places. And thus multiply the whole upper number by the whole lower number.

Vogel [1963, p. 46] interpreted the procedure of the above problem as follows:

2326	428326	428326	492226
214	214	214	214

496486	497764
214	

Kūshyār ibn Labbān also described the same method in *Kitāb fī Usūl Hisāb al-Hind* (c. 1000) [Levey & Petruck 1965, pp. 52–56]. He used the example 325 multiplied by 243.

3.4 Division

3.4.1 The method in *Sun Zi suanjing*

Just as in multiplication where its operations are based on the position of the multiplier relative to the multiplicand, so in division the operations are based on the placing of the divisor relative to the dividend. The initial position of the divisor relative to the dividend determines the place value of the first digit (from the left) of the quotient. In describing the procedure for division, *Sun Zi suanjing* begins by showing how to place the divisor in relation to the dividend.[2] The passage (p. 194) is as follows:

> In the common method of division (*fan chu zhi fa* 凡除之法), this is the reverse of multiplication. The dividend (*cheng de* 乘得 lit. product) occupies the middle position and the quotient (*chu de* 除得) is placed above it. Suppose 6 is the divisor (*fa* 法) and 100 is the dividend (*shi* 實). When 6 divides 100, it advances (*jin* 進) two places [to the left] so that it is directly below the hundreds. This implies the division of 1 by 6. In this case, the divisor (*fa*) is greater than the dividend (*shi*), so division is not possible. Therefore shift (*tui* 退) [6 to the right] so that it is below the tens. Using the divisor (*fa*) to remove the dividend (*shi*), one six [is 6] and 100 is reduced to 40, thus showing that division is possible. If the divisor (*fa*) is less than [that part of] the dividend [above it] (*shi*), it should then stay below the hundreds and should not be shifted. It follows that if the units of the divisor (*fa*) are below the tens [of the dividend], the place value [of the digit of the quotient] is tens; if they are below the hundreds, the place value [of the digit of the quotient] is hundreds.[3] The rest of the method is the same as

---

[2] *Xiahou Yang suanjing* 夏侯陽算經 has a similar explanation [Qian ed. 1963, p. 558]. For a translation of this, see [Lam 1987, pp. 373–374].

[3] To elucidate the concept of blank spaces, Li Chunfeng 李淳風 inserted the following: "If [the numeral of] the upper position has an empty space, this means that the divisor has been shifted two places [to the right]". This is the only comment of Li Chunfeng in the book. See p. 30.

multiplication. As for the remainder of the dividend (*shi*), this is assigned to the divisor (*yi fa ming zhi* 以法命之) such that the divisor (*fa*) is called the denominator (*mu* 母) and the remaining dividend (*shi*) the numerator (*zi* 子).

The passage explains that in dividing 100 by 6, the numerals are initially displayed as in [i]. The divisor 6 is then shifted to the extreme left [ii]. Division is not possible since 6 cannot divide 1. The divisor 6 is shifted one place to the right [iii]. Since 6 is below the tens of 100, the first digit of the quotient is in the tens place and should be placed above the tens of the dividend. One six is 6, so 1 is the quotient and 100 is reduced to 40 [iv]. The rest of the working is not explained explicitly, but it is easy to proceed. The divisor 6 is shifted to the right by one place [v]. Six sixes are 36, the quotient 6 is in the units place and 40 is reduced to 4 [vi]. The remainder 4 is called *zi* 子 (numerator, lit. son) and the divisor 6 is called *mu* 母 (denominator, lit. mother). The notation in [vi] is equivalent to our notation $16\frac{4}{6}$.

Converted to Hindu-Arabic numerals, the above operations appear as follows:

			1	1	16
100	100	100	40	40	4
6	6	6	6	6	6
[i]	[ii]	[iii]	[iv]	[v]	[vi]

We draw attention here to two very frequently used technical words in Chinese mathematics. They are *shi* 實 and *fa* 法. In the above passage concerning division, we have translated them as "dividend" and "divisor" respectively. However they have also been used in the description of other methods such as the extraction of roots and the solution of polynomial equations. In such cases, their meanings would be different. Besides having a general connotation, *shi* and *fa* to the computer are also associated with numerals occupying certain positions on a counting board. They are therefore associated with the changes that these numerals undergo during the different stages of the method. In the above example, the numerals in the middle position, i.e., 100, 40 and 4, are referred to as *shi*, and the numeral 6 in the lower position is referred to as *fa*.

When the operation of dividing finally leaves a remainder in the *shi* which is smaller than the divisor, this remainder and the divisor constitute a fraction which is expressed by the technical phrase *yi fa ming zhi* 以法命之 meaning "assign [the remainder] to the divisor". In a proper fraction, the numerator is usually referred to as *zi* 子 which means "son" and the denominator as *mu* 母 which means "mother".

The quotient may have a zero among its digits, which is a blank space in the rod numeral notation. Li Chunfeng noted that if this was the case (fn. 3), the divisor would then have been shifted two places to the right, the blank space of the quotient being the result of the first shift to the right. We illustrate this in the following example where 102 is derived when 612 is divided by 6 as shown below:

1	10	102
612	12	
6	6	6

What Li Chunfeng said is clearer when depicted in rod numerals as shown below:

Ch. 1, Prob. 2, which is quoted below, illustrates clearly and in detail the stages of the operation of division.

> If 6,561 is divided among 9 persons, find how much each gets.
> Answer: 729.
> Method: First set 6,561 in the middle position to be the *shi* 實 (dividend). Below it, set 9 persons to be the *fa* 法 (divisor) [i]. Put down 700 in the upper position [ii]. The upper 7 calls the lower 9:

seven nines are 63, so remove 6,300 from the numeral in the middle position [iii]. Shift the numeral in the lower position one place [to the right] and put down 20 in the upper position [iv]. The upper 2 calls the lower 9: two nines are 18, so remove 180 from the numeral in the middle position [v]. Once again shift the numeral in the lower position one place [to the right], and put down 9 in the upper position [vi]. The upper 9 calls the lower 9: nine nines are 81, so remove 81 from the numeral in the middle position [vii]. There is now no numeral in the middle position. Put away the numeral in the lower position. The result in the upper position is what each person gets [viii].

[i]	[ii]	[iii]	[iv]

[v]	[vi]	[vii]	[viii]

The above steps are shown below in Hindu-Arabic numerals.

	7	7	72
6561	6561	261	261
9	9	9	9
[i]	[ii]	[iii]	[iv]

72	729	729	729
81	81		
9	9	9	
[v]	[vi]	[vii]	[viii]

We now summarise the Chinese method of division.

1. The dividend and the divisor are set on the board with the divisor directly below the dividend.
2. The divisor is shifted to the left such that its first digit from the left is aligned with the first digit of the dividend.

3. If that part of the dividend immediately above the divisor is numerically larger or equal to the divisor then division is possible, and the first digit of the quotient is obtained and placed above the dividend in the same column as the units of the divisor. Its place value is the same as the place value of the digit of the dividend immediately below it. If that part of the dividend immediately above the divisor is numerically smaller than the divisor, division is not possible, and the divisor is shifted one place to the right. After the divisor has been shifted, it is now possible to proceed with division and when the first digit of the quotient is obtained, it is placed above the dividend in the same column as the units of the divisor. The place value of this digit is the same as the place value of the digit of the dividend immediately below it.

4. The digit of the quotient multiplies each digit of the divisor commencing from the left. Each product is placed such that the units occupy the same column as the digit that has been multiplied and the tens are to the left of the units. These are subtracted from the digits of the dividend occupying the same columns.

5. When subtraction is completed, the divisor is shifted one place to the right. The same procedure is repeated with the remaining dividend until there is either no remainder or the remainder is less than the divisor.

6. The place values of the digits of the quotient in the first row correspond to the place values of the digits of the dividend in the second row. These place values remain stationary and are thus fixed. The digits of the divisor in the third row do not correspond to these place values because the divisor moves to the right stage by stage.

As the procedures of multiplication and division followed a repetitive algorithmic pattern, they could be applied without difficulty to numerals however large. It is therefore not surprising that at a very early period in Chinese history, there were numerous instances of the multiplication and division of large numerals. In *Sun Zi suanjing* (p. 200) we find the following statement:

> Multiply 708,588 by 531,441 to obtain 376,572,715,308. When this is divided among 354,294 persons, each person gets 1,062,882.

The invention of the methods of multiplication and division (together with addition and subtraction) was possible because of the conceptual basis of the rod numeral system, which used a place value notation with ten as base. Undoubtedly these arithmetical procedures played an important role in the advancement of all fields that required calculation. Their remarkable simplicity places them amongst the greatest achievements of mankind.

3.4.2   The earliest method in Islam

The earliest method of division using Hindu-Arabic numerals as recorded in Arabic books is identical with the Chinese method. This is most remarkable as there is a difference of a few centuries between these books and *Sun Zi suanjing*, and, furthermore, their media of computation were very different. The text of Cambridge University Library Ms. Ii.vi.5 based on al-Khwārizmī's arithmetic has a general description of division followed by an illustration of two examples, 46,468 divided by 324 and 1,800 divided by 200 (fol. 108r–109v). We quote below the English translation by Crossley & Henry [1990, pp. 117–118] on the passage concerning division in general.

> In division on the other hand put the number that you wish to divide in its places; next put the number itself by which you wish to divide under it. And let the last place of the number by which you are dividing be under the last place of the upper number that you are dividing. Moreover, if the number that is the last place of the upper number that you wish to divide is less than that which is the last place of the lower number by which you are dividing, retract the place itself toward the right, until the number of the upper number is greater, i.e., put the last place of the lower number by which you are dividing under the second place that comes after the last place of the upper number. After this consider the first place of the number by which you wish to divide and put in its column above the upper number that you are dividing or beneath it in its column some number which, when you have multiplied it by the last place of the lower number by which you are dividing, will be the same as that number that was in the upper place or close to it, provided it is less than it. When you know it, multiply it by the last place of the lower number and subtract what results to you from the multiplication from that which is above it, from the upper number that is being divided. Once again multiply it

by the second place that comes after the last place toward the right, and subtract it from that which is above it and proceed in the division just as you proceeded in the beginning of the book, when you wished to subtract some number from another number; and likewise proceed until you multiply it by all the places of the lower number by which you are dividing. After this move all the places of the lower number, by which you are dividing, one place toward the right......

The description reveals that the procedure is the same as the Chinese method. The phrase "the last place of a number" refers to the first digit of the number from the left.

Besides translating the Latin version of al-Khwārizmī' s arithmetic into German, Vogel [1963, p. 47] added a commentary in which he summarised the description of the division of 46,468 by 324 in the following steps:

1	1	14	14	14
46468	14068	14068	2068	1268
324	324	324	324	324
14	143	143	143	143
1108	1108	208	148	136
324	324	324	324	324

The format is identical with the Chinese method down to the last step which shows the answer as $143\frac{136}{324}$.

Al-Uqlīdisī gave numerous examples in his arithmetic book to illustrate the same procedure for the earliest method of division. The first of these is 144 divided by 6, which is described in the following manner [Saidan 1978, p. 55]:

The working is to assume the six below the four and seek something to multiply by six so as to cover the most possible of the fourteen, the one after the four being 10 with respect to the six. The arrangement is 144. We find that to be 2. Having found out what to multiply by 6,
   6
we insert it above the six and the four, as follows: 2 . Now we
           144
           6

multiply the two by the six, saying: ' 2 by 6 : 12' . We drop away the twelve from the fourteen, leaving 2 in the place of four.
We shift the six setting it below the four. We get 2 . Now we seek

$$\begin{array}{c} 24 \\ 6 \end{array}$$

something to multiply by 6 and get 24. We find that four. So we insert four above four in the form: 24. Now we say: ' Six by four is 24' ,

$$\begin{array}{c} 24 \\ 6 \end{array}$$

which exhausts the twenty-four, leaving nothing. The quotient is on top, which is 24. It is the answer.

In Kūshyār ibn Labbān' s book, the method of 5,625 divided by 243 is described in detail and illustrated in the following stages [Levey & Petruck 1965, pp. 58–60]:

	2	2	23	23
5625	5625	765	765	36
243	243	243	243	243
[i]	[ii]	[iii]	[iv]	[v]

We quote from it an example of the description of its procedure from stage [ii] to [iii].

> We multiply it [that is, 2 in the first row] by the lowest 2 to get 4 which is subtracted from the 5 that is above the lowest 2. Also, we multiply by the lower 4 and subtract it from what is above the 4. Again, we multiply by the lower 3 and subtract it from what is above the 3. Then we shift the lower orders one place.

The above examples illustrate the earliest method on division used in Islam. Later methods, see [Saidan 1978, pp. 270–275], showed a deviation in procedures which made them different from the original, apart from the basic use of place values. Since arithmetic procedures are mere conventions which evolve through experimentations in which socio-cultural backgrounds are inextricably integrated, why are the earliest methods of multiplication and division in Islam identical with the Chinese methods? This will be discussed in Sect. 9.

3.5   Addition and subtraction

The methods of addition and subtraction using rod numerals are not explained explicitly in *Sun Zi suanjing*. These operations were probably considered too simple and too well known to require written explanations. As we have seen earlier, the procedure for multiplication involves addition, and the procedure for division involves subtraction. From these two procedures, it is not difficult to infer the addition and subtraction procedures.

In the case of multiplication, the operation commences from left to right. Addition is involved here and this is also performed in the same direction from left to right. Therefore, in the procedure of adding two numerals, we can assume that one numeral is placed below the other such that the place values of their digits correspond; furthermore the operation starts from left to right. In stages [iv] and [v] of the description for the example 81 multiplied by 81 (p. 59), 80 is added to 6,480 so that the latter changes to 6,560. From this, we deduce, that when the lower digit is added to the digit immediately above, the upper digit is changed to the sum of the two digits. If the sum exceeds nine, one ten is added to the digit on the left. (See also Sect. 3.2.)

We summarise our conclusions on the procedure of adding two numerals as follows:

1. Set one numeral below the other, units under units, tens under tens, etc.
2. Commencing from the left, add the lower digit to the upper digit. In other words, the rods of the lower digit are added to the rods of the digit immediately above, so that the upper digit is changed to the sum of the two digits. If the sum exceeds nine, the ten (represented by one rod) is added to the digit on the left.
3. In this manner, the upper numeral gradually changes into the sum, and the lower numeral disappears.

Let us illustrate the procedure in the example, 2,546 + 371. On the counting board, the numerals are set as follows:

Commencing from the left, 3 is added to 5 and the sum 8 takes the place of 5, so that the upper numeral becomes 2,846 [ii]. Next 7 is added to 4 and the sum is 11; the 4 and 8 of the upper numeral are then changed to 1 and 9 respectively. The upper numeral is now 2,916 [iii]. Lastly 1 is added to 6 and the result is 2,917 [iv]. The rod digits are added in the manner described in Sect. 3.2.

= ‖‖ ≣ T	= π ≣ T	= Ⅲ − T	= Ⅲ − π
‖ ⊥ │	⊥ │	│	
[i]	[ii]	[iii]	[iv]

The above steps are shown below in terms of Hindu-Arabic numerals.

2546	2846	2916	2917
371	71	1	
[i]	[ii]	[iii]	[iv]

Division is the reverse process of multiplication, and subtraction is the reverse of addition. Just as we have inferred the process of addition from that of multiplication, we can also infer the process of subtraction from that of division. To do this, we note in particular, stages [iii] and [v] of the example 6,561 divided by 9 (p. 66). We now write down the procedure for the subtraction of one numeral from another in the following stages.

1. Set the subtrahend below the minuend, units under units, tens under tens, etc.
2. Commencing from the left, subtract the lower digit from the upper digit. If the upper digit exceeds the lower digit, the subtraction procedure removes the lower digit and changes the upper digit into the difference. If the upper digit is less than the lower digit, one ten taken from the digit on its left is added to it to enable subtraction to take place.
3. In this manner, the upper numeral gradually changes into the difference, and the lower numeral disappears.

The procedure of 2917−371 is shown below. As in addition, the subtraction of rod digits is performed according to the rules governing vertical and horizontal rods (Sect. 3.2).

$$= \text{Ⅲ} - \text{Ⲧ} \qquad = \text{Ⲧ} - \text{Ⲧ} \qquad = \text{ⅢⅠ} \equiv \text{Ⲧ} \qquad = \text{ⅢⅠ} \equiv \text{Ⲧ}$$
$$\text{Ⅲ} \perp \text{Ⅰ} \qquad\qquad \perp \text{Ⅰ} \qquad\qquad \text{Ⅰ}$$

The above steps in Hindu-Arabic numerals are as follows:

2917	2617	2547	2546
371	71	1	

If we compare these methods of addition and subtraction with the earliest methods described in the three Arabic texts mentioned above, we discover once again that the methods are the same [Saidan 1978, pp. 46−48; Levey & Petruck 1965, pp. 48−52; Vogel 1963, pp. 17−21, 45−46; Crossley & Henry 1990, pp. 114−115].

3.6   The multiplication table

The methods of multiplication and division required knowledge of the multiplication table. The inception of the multiplication table in China could be traced to remote antiquity. It was likely that when "calculation" (*shu* 數) was instituted as one of the six arts (*liu yi* 六藝) in the education program for the imperial princes of Zhou 周 in the 8th century BC, the multiplication table could have formed an integral part in the rudiments of counting.[4] What is certain is that by the 7th century BC, the multiplication table was popularly learnt and widely circulated. This is evident in an anecdote recorded in Han Ying's 韓嬰 *Han shi wai chuan* 韓詩外傳 (The exposition

---

[4] The six arts referred to in *Zhou li* 周禮 (Records of the rites of Zhou) were propriety, music, archery, charioteership, calligraphy and mathematics. Liu Caonan 劉操南 even thought that the item on mathematics meant specifically calculation using the multiplication table. (See [Needham 1959, p. 25].)

of poetry by Han Ying) of early Han.[5] It told of Qi Huangong's 齊恒公 (Duke Huan of the state of Qi, 685–643 BC) extensive search for talents to serve the state. He waited for a long time and no one applied, until one day someone approached him with the multiplication table. Qi Huangong was sceptical of the person's motive. The person told him that if he considered someone with only the knowledge of the multiplication table, he would not be short of talents coming forward to present themselves. True enough, as the person was warmly received, a host of other talents began to appear.

The story confirms that the multiplication table was in widespread use during the powerful reign of Qi Huangong. The popularity of the table was

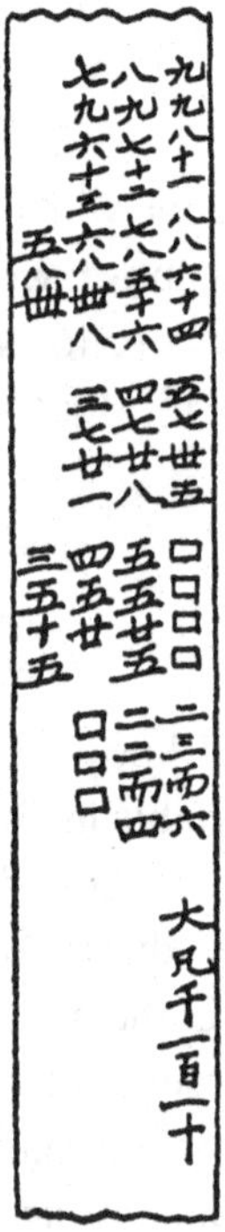

		7	8	9
		9	9	9
		63	72	81
	5	6	7	8
	8	8	8	8
	40	48	56	64
		3	4	5
		7	7	7
		21	28	35
	3	4	5	
	5	5	5	
	15	20	25	
			2	2
			2	3
			are	are
			4	6
				total
				1110

Fig. 3.1 A reproduction of the legible characters of a multiplication table tablet unearthed at Liu Sha 流沙 and its translation.

[5] This anecdote was commonly quoted by historians of mathematics in China. See, for example, [Li Yan 1954, p. 12; Qian 1964, p. 10].

also due to the mnemonic nature of its composition. Unlike the modern multiplication table which enumerates from one one in ascending order, the ancient Chinese multiplication table started from nine nines in descending order. The numbers were arranged in columns and rows without using any auxiliary words such that the first two numbers were the multiplicand and multiplier, and the number which followed was the product.

The earliest extant examples of the multiplication table were found on the remnants of the Han wooden and bamboo strips unearthed in the north western region of China towards the end of the 19th century. Because of their perishable nature, certain portions of the strips were destroyed, rendering the writing on them irretrievably lost. A wooden strip found in Liu Sha 流沙 had the multiplication table [Li Yan & Du Shiran 1963, p. 20]. The strip measured 260 mm in length and 24 mm in width. It was indicated on the strip that there were altogether 1,110 individual characters, which included the numbers written in characters. Li Yan [1954, pp. 16–18], who quoted from studies of the strip by Luo Zhenyu 羅振玉 and Wang Guowei 王國維, confirmed that the table ended with two twos. However since the lower portion of the strip was partially putrefied, the table was tantalizingly incomplete. The legible characters of the multiplication table are reproduced in Fig. 3.1 with their translation on the right.

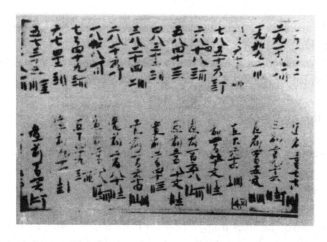

Fig. 3.2 A section of the multiplication table in *Li cheng suanfa* 立成算法 of the Dunhuang scroll.

In ancient and medieval times, the multiplication table was commonly known as the "nine nines song", probably because a learner of mathematics had to recite the numbers in a singing manner to commit the table to memory. As such, the table could be easily remembered and transmitted from one person to another. As it was widely popularised, examples of multiplying one number by another in the "nine nines song" style were found in many classical texts [Li Yan 1954, pp. 14–16]. The table was also carefully copied on scrolls and preserved in the Dunhuang Caves during the Tang dynasty [Libbrecht 1982, pp. 212 & 218]. This type of multiplication table has a special feature: each product displayed is repeated in rod numerals. According to Li Yan & Du Shiran [1987, p. 15], this is the earliest extant material displaying rod numerals in a written form. (See Fig. 3.2.) From all these available sources, it is learnt that during the first and second century AD, the table began with nine nines and ended with two twos. It was extended to one one probably from the 5th century AD. Nevertheless, by the 13th century, the table began to be presented from one one to nine nines in ascending order as it is today.

The first problem of *Sun Zi suanjing* informs the reader that nine nines are 81, and 81 multiplied by itself is 6,561. The second problem involves the division of 6,561 by 9. This is followed by the same kind of exercises applied to each successive recitation of the multiplication table, namely (p. 196):

> Eight nines are 72; multiply this by itself to obtain 5,184. When this is divided among 8 persons, each person gets 648.
> Seven nines are 63; multiply this by itself to obtain 3,969. When this is divided among 7 persons, each person gets 567.

The nine times table ends with

> One nine is 9; multiply this by itself to obtain 81. One person gets 81.

After this, we are told that

> [The sum of the nines from] nine nines listed above is 405; multiply this by itself to obtain 164,025. When this is divided among 9 persons, each person gets 18,225.

The same format is applied to the eight times table, that is, eight eights, seven eights, etc. ending with the sum of the eights. The pattern is repeated successively till "one one is 1". A grand total is then obtained which is derived in the following manner:

$$
\begin{aligned}
9\times9 + 8\times9 + 7\times9 + 6\times9 + 5\times9 + 4\times9 + 3\times9 + 2\times9 + 1\times9 &= 405 \\
8\times8 + 7\times8 + 6\times8 + 5\times8 + 4\times8 + 3\times8 + 2\times8 + 1\times8 &= 288 \\
7\times7 + 6\times7 + 5\times7 + 4\times7 + 3\times7 + 2\times7 + 1\times7 &= 196 \\
6\times6 + 5\times6 + 4\times6 + 3\times6 + 2\times6 + 1\times6 &= 126 \\
5\times5 + 4\times5 + 3\times5 + 2\times5 + 1\times5 &= 75 \\
4\times4 + 3\times4 + 2\times4 + 1\times4 &= 40 \\
3\times3 + 2\times3 + 1\times3 &= 18 \\
2\times2 + 1\times2 &= 6 \\
1\times1 &= 1
\end{aligned}
$$

Grand total   1155

Lastly we are asked to multiply 1,155 by itself to obtain 1,334,025, and this is divided among 9 persons (p. 200).

3.7   Conclusion

A primitive way to count and to do simple additions and subtractions was through the manipulation of bamboo sticks. Commencing with them, the ancient Chinese evolved the rod numeral system with its three significant features. The first feature showed that for digits 6 to 9, five units were grouped together to be represented by one rod. A distinction had to be made between the value of a vertical rod and that of a horizontal one, for one type of rod had to represent one unit and the other type five units. The next feature of the rod system was the use of a place value notation with ten as base. The third feature was that the value of a vertical rod alternated between one unit and five units in successive digits of a numeral commencing from the right, and the value of a horizontal rod alternated between five units and one unit.

The addition of a small number to a rod numeral on the board or the subtraction of a small number from that numeral would involve the manipulation of rods on that numeral. This developed into the addition and subtraction methods discussed in Sect. 3.5, where the numeral in the upper row was changed stage by stage from left to right simultaneously as the numeral in the lower row was removed digit by digit from left to right.

Similar to the methods of addition and subtraction, the method of multiplication commenced from the left. The placement of the multiplier relative to the multiplicand was important, for the product had to be placed in relation to both numerals. The procedure involved the removal of each digit of the multiplicand after it had been multiplied and the shifting of the multiplier step by step to the right. The process of division also required the learning of corresponding techniques. The divisor had to be placed below the dividend according to certain rules, and the quotient had to be placed in relation to the position of the divisor and the dividend. As the dividend diminished, the divisor had to be shifted step by step to the right.

The earliest methods of adding, subtracting, multiplying and dividing with Hindu-Arabic numerals as recorded by the Arabs were remarkably the same as those of the Chinese methods.[6] This was so despite the fact that the Hindu-Arabic numeral system was a written one and the rod numeral system was not. Moreover there were details in the description of the operations, such as the removal of the digits of a numeral or the shifting of a numeral from place to place, which showed that they were natural actions from a rod and board system but were not conducive to a written system.

There are numerous ways of performing the four fundamental operations of arithmetic. In fact in the Arab countries and Europe, after the Hindu-Arabic numeral system and the above arithmetic methods were understood and learned, other methods of operation began to appear. These were more conducive to a written numeral system. The Chinese themselves using rod numerals did not adhere only to the methods that we have described. For instance, Yang Hui 楊輝 in his *Cheng chu tong bian suan bao* 乘除通變算寶 (Precious reckoner for variations of multiplication and division), gave a variety of methods on multiplication and division (see [Lam 1977, pp. 5−82]). The fact that the earliest methods in Islam and those of *Sun Zi suanjing* are the same is a remarkable phenomenon.

---

[6] These methods can also be found amongst the earliest arithmetic texts in Europe. For example, A.J.E.M. Smeur's analysis of the multiplication and division methods of the first arithmetic printed in the Netherlands, *Arithmeticae summa tripartita* by Georgij de Hungaria [1499, pp. 16−19], showed them to be the same as the Chinese methods.

chapter 4

The Common Fraction

4.1 Concept, notation and rules

The rod numeral system provided the means through which an efficient and concise method of division was devised. The remainder in the division process led to the crystalization of the concept of what we call a common fraction. In such a manner the notation of a fraction was invented. This concept is clearly described in *Sun Zi suanjing* (pp. 63–65 & p. 194) and *Xiahou Yang suanjing* [Qian ed. 1963, p. 558] in terms of the non-zero remainder of the division process and the divisor. The remainder called *zi* 子 which means "son", is "assigned to the divisor" *yi fa ming zhi* 以法命之; and the divisor is called *mu* 母 which means "mother". We have translated *zi* as "numerator" and *mu* as "denominator".

The notation of the common fraction using rods was automatically assumed from the display of the final stage in the division procedure. This consisted of *zi* 子 (numerator) placed immediately above *mu* 母 (denominator). Let us illustrate this by taking the example, 100 divided by 6, which gave 16 as quotient and 4 as remainder (see p. 64); the last stage displayed on the counting board was:

This stood for $16\frac{4}{6}$, the notation for the fraction, $\frac{4}{6}$, being $\frac{||||}{\top}$. Hence a fraction expressed with rod numerals has a very simple form: the numeral of the fractional part is placed above the numeral representing the whole.

In a written statement a fraction is expressed in words so that $\frac{4}{6}$ is written as 六分之四 (*liu fen zhi si*), which means "4 parts out of 6". The written expression of a fraction and the concise notation of a fraction through rod numerals are therefore two very different representations. The subject of fractions was developed through the rod numeral notation: it was through experimentations with fractions on the counting board that the ancient Chinese mathematicians were the earliest to evolve an extensive and systematic study on fractions. When the derivations and results from the counting board were recorded in treatises, fractions were written in characters which followed their rhetoric expressions. The stage by stage performances with rods were converted to brief descriptions in words. It is therefore not surprising if a person reading an ancient treatise on fractions is unable to follow the subject fully — unless he envisages the computational stages on the counting board.

In our translation of *Sun Zi suanjing*, we have used our present notation to stand for the written expression of fraction instead of translating it literally. For example, 六分之四 *liu fen zhi si*, which literally means "4 parts out of 6" is translated as $\frac{4}{6}$. The notation, $\frac{4}{6}$, is the same as in rod numerals except for the horizontal line that separates the numerator from the denominator.[1]

The variety of problems on fractions in *Jiu zhang suanshu*, (the main ones being Ch. 1, Probs. 5–24; Ch. 4, Probs. 3, 4, 6–10) manifests the depth of knowledge that the ancient mathematicians possessed. The topics discussed include the reduction of a fraction, the addition, subtraction, multiplication, division and averaging of fractions, as well as improper fractions. The recorded methods were extremely terse; and this had prompted Liu Hui to explain some of the passages.

---

[1] The notation of a fraction in Hindu-Arabic numerals was originally of the same form as the notation in rod numerals without a horizontal line. The extra horizontal line separating the numerator from the denominator of the present form was introduced probably in the 12th century [Cajori 1928, p. 269].

There are only four problems on fractions in *Sun Zi suanjing* (Ch. 2, Probs. 1–4) and all of them are from *Jiu zhang suanshu* (Ch. 1, Probs. 5, 7, 10 & 15). In the latter book the four problems are stated with a brief statement of the general method. In *Sun Zi suanjing* more detailed procedures for solving them are given — these have provided an important insight into how fractions were computed with rods. Each problem illustrates a particular rule, which we shall now analyse.

4.2 The reduction of a fraction

The first problem concerns the reduction of a fraction. If the division of one number by another results in a remainder, the fraction that is formed from this remainder and the divisor sometimes needs to be reduced to its lowest term. The procedure of reducing a fraction is called *yue fen* 約分. The following passage contains Liu Hui's comments on the subject [Qian ed. 1963, pp. 94–95].

> On [the subject of] *yue fen* when a quantity is not a whole, we speak of it in terms of parts. The numbers associated with the parts are complex and difficult for application. Take "two parts out of four" as an example; in complex terms, it is also "four parts out of eight", and in simplified terms, it is "one part out of two". Although the expressions are different, they are nevertheless of the same quantity. The *fa* 法 (divisor) and *shi* 實 (dividend) are mutually manipulated when their differences are taken, and in this manner all the parts are first dealt with.

We now examine the first problem on fractions from *Sun Zi suanjing* (Ch. 2, Prob. 1):

> Now there is a fraction $\frac{12}{18}$; reduce (*yue* 約) it to find its [simplest] form.
> Answer: $\frac{2}{3}$.
> Method: Put down 18 in the lower position and 12 in the upper position. Set the numerals in two other positions for the purpose of subtracting the smaller from the larger to derive 6 as the *deng shu* 等數 (greatest common divisor lit. equal number). Use it as a divisor (*fa* 法) for reducing (*yue zhi* 約之) [the fraction] to obtain the answer.

The problem is concerned with the reduction of $\frac{12}{18}$, and the first stage is to derive the *deng shu* 等數 (greatest common divisor) of 12 and 18. This involves a series of subtractions commencing with the denominator and numerator of the given fraction. The subtractions here are $18 - 12 = 6$; $12 - 6 = 6$. The series of subtractions ends when there is a pair of equal numbers in the subtrahend and the difference; in this case, the number is 6. After the *deng shu* is found, it is used to divide the numerator and denominator of the given fraction, resulting in the fraction being reduced to its lowest term.

The above series of subtractions would probably be performed on the counting board in the following manner: first, 18 is placed in the lower position with 12 above [i]; next, 12 is subtracted from 18, and the difference 6 replaces the numeral 18 [ii]; lastly, 6 is subtracted from 12 and the difference 6 replaces 12 [iii]. The pair of equal numerals indicates the *deng shu*.

[i] [ii] [iii]

The procedure of diminishing the denominator and numerator of a fraction through a series of subtractions is described in the following passage of *Jiu zhang suanshu* [Qian ed. 1963, p. 95]:

> If both [the denominator and numerator] can be halved then halve them. When both cannot be halved, set down the numerals of the denominator and numerator, and subtract the smaller from the larger. Continue to diminish mutually through subtractions (*geng xiang jian sun* 更相減損) to seek a pair of equal numbers (*deng shu* 等數). Use [this number called] the *deng shu* to reduce [the fraction].

This often quoted description is the ancient Chinese method of finding the greatest common divisor of two numbers. The term, *deng shu*, whose literal meaning is "equal numbers", is their technical phrase for "greatest common divisor". Liu Hui offered the following explanation on why the *deng shu* was used for reducing a fraction [Qian ed. 1963, p. 95]:

To reduce by using the *deng shu* means to divide [the denominator and numerator] by it. As the mutually subtracted numbers are all multiples (*chong die* 重疊) of the *deng shu*, this is the reason why the *deng shu* is used to reduce [the fraction].

4.3 The addition and subtraction of fractions

The addition of fractions is called *he fen* 合分 and this procedure is illustrated in *Sun Zi suanjing* Ch. 2, Prob. 2:

Now there are fractions $\frac{1}{3}$ and $\frac{2}{5}$, find their sum (*he* 合).
Answer: $\frac{11}{15}$.
Method: Put down denominators (*fen* 分) 3 and 5 on the right and numerators (*zhi* 之) 1 and 2 on the left [i]. Multiply a numerator and the other denominator (*mu hu cheng zi* 母互乘子) to obtain 6 for the fraction $\frac{2}{5}$ and 5 for the fraction $\frac{1}{3}$ [ii]. Add to obtain 11, which becomes the dividend (*shi* 實). Multiply the two denominators on the right to give 15, which becomes the divisor (*fa* 法). [The dividend] is less than the divisor, so assign it to the divisor (*yi fa ming zhi* 以法命之) to get the answer [iii].

[i]	[ii]	[iii]

The above stages on the counting board are shown below in terms of Hindu-Arabic numerals.

1 3	5	11
2 5	6	15
[i]	[ii]	[iii]

The method manifested that fractions were treated through the notion of whole numerals. Since the four basic arithmetic operations of the rod numerals were known, such operations were extended to the fractions. Certain arrangements of numerals on the board were necessary: the denominators of the fractions were displayed vertically on the right and the corresponding numerators on their left. This sort of arrangement led naturally to the conception of the cross-multiplication of numerals, (see [i] above). The Chinese called such a performance on the counting board *mu hu cheng zhi* 母互乘之 meaning "multiply mutually each numerator and the other denominator(s)".

The procedure for the subtraction of fractions (*jian fen* 減分) is similar to that for the addition of fractions as shown in Ch. 2, Prob. 3.

> Now there is a fraction $\frac{8}{9}$ from which $\frac{1}{5}$ is subtracted (*jian* 減). Find the remainder.
>
> Answer: $\frac{31}{45}$.
>
> Method: Put down denominators 9 and 5 on the right and numerators 8 and 1 on the left [i]. Multiply a numerator and the other denominator to obtain 9 for the fraction $\frac{1}{5}$ and 40 for the fraction $\frac{8}{9}$ [ii]. Subtract the smaller from the larger to obtain a remainder, 31, which becomes the dividend (*shi*). Multiply the denominators to give 45, which becomes the divisor (*fa*). [The dividend] is less than the divisor, so assign it to the divisor (*yi fa ming zhi* 以法命之) to get the answer [iii].

𝍪 𝍫	≡	≡ ∣	
∣ ⦀⦀	𝍫	≡ ⦀⦀⦀	
[i]	[ii]	[iii]	

The above translated into Hindu-Arabic numerals is as follows:

8	9	40	31
1	5	9	45
[i]		[ii]	[iii]

The addition of fractions is described in *Jiu zhang suanshu* [Qian ed. 1963, p. 95–96] in the following manner:

> Multiply mutually each numerator and the other denominator(s) (*mu hu cheng zi*), and let the sum [of the products] be the dividend (*shi*). Multiply the denominators and let [the product] be the divisor (*fa*). Divide (*shi ru fa er yi* 實如法而一, lit. divide the dividend by the divisor). When the remaining [dividend] is less than the divisor, assign it to the divisor (*yi fa ming zhi*) [to form a fraction].

The rule stated in this brief passage, which contains only 26 characters, can be applied to the addition of more than two fractions. If there are, say, three fractions, $\frac{a}{x}, \frac{b}{y}$ and $\frac{c}{z}$, then the first sentence "multiply mutually each numerator and the other denominators" refers to the products ayz, bxz and cxy. This process and the other process of multiplying all the denominators together, namely, xyz, were explained by Liu Hui in the following passage [Qian ed. 1963, p. 96]:

> Multiplying mutually each numerator and the other denominator(s) is called *qi* 齊, and multiplying all the denominators together is called *tong* 同. The *tong* process attains a common denominator, and the *qi* process of multiplying each numerator and the other denominator(s) is to preserve the original value [of each fraction].

Jiu zhang suanshu has a description for the subtraction of two fractions which is similar to the rule for the addition of fractions [Qian ed. 1963, p. 96]:

> Multiply mutually each numerator and the other denominator; take away the smaller quantity from the larger and let the remainder be the dividend (*shi*). Multiply the denominators and let [the product] be the divisor (*fa*). Divide (*shi ru fa er yi* 實如法而一, lit. divide the dividend by the divisor).

This is followed by the rule for the comparison of fractions which is called *ke fen* 課分 [Qian ed. 1963, p. 97]. The method is the same as that for the subtraction of fractions. What differs in the meaning of the two processes, as explained by Liu Hui, is that, in the latter we are required to find the remainder, while in the former we are required to know the excess of one fraction over the other.

4.4 The averaging of fractions

Like the other three problems, the solution of the fourth problem in Chapter 2 of *Sun Zi suanjing* provides a valuable insight into how fractions were computed with rods. The problem demonstrates the rule of averaging fractions called *ping fen* 平分, and is stated as follows:

Now there are fractions $\frac{1}{3}$, $\frac{2}{3}$ and $\frac{3}{4}$. Find the amounts to be subtracted from the larger [fractions] and that to be added to the smallest [in order to obtain] the average value (*ping* 平), and find this value.

Answer: Subtract 2 [twelfths] from $\frac{3}{4}$ and 1 [twelfth] from $\frac{2}{3}$; the sum [of the subtrahends] is added to $\frac{1}{3}$. Each gives the average value of $\frac{7}{12}$.

Method: Put down denominators 3, 3 and 4 on the right and numerators 1, 2 and 3 on the left [i]. Multiply a numerator and the other denominators [ii]; add [the products] to obtain 63 and put this on the right calling it *ping shi* 平實 (lit. average dividend) [iii]. Multiply the denominators to obtain 36, which becomes the divisor (*fa*) [iv]. Use 3, which is the number of the displayed [fractions], to multiply [each of the products] before their addition [v] and the divisor (*fa*) [vi]. 9 is derived as the *deng shu* 等數 (greatest common divisor) and this reduces [the fractions] [vii]. Subtract 2 [twelfths] from $\frac{3}{4}$ and 1 [twelfth] from $\frac{2}{3}$ [viii]. The sum [of the subtrahends] is added to $\frac{1}{3}$ [ix]. Each gives the average value of $\frac{7}{12}$.

We analyse the working of the above problem first in terms of Hindu-Arabic numerals:

1　3		$1 \times 3 \times 4 = 12$	$12 + 24 + 27 = 63$
2　3		$2 \times 3 \times 4 = 24$	
3　4		$3 \times 3 \times 3 = 27$	
[i]		[ii]	[iii]

$3 \times 3 \times 4 = 36$	$12 \times 3 = 36$	$36 \times 3 = 108$
	$24 \times 3 = 72$	
	$27 \times 3 = 81$	
[iv]	[v]	[vi]

$\frac{18}{108} = \frac{2}{12}$	$\frac{3}{4} - \frac{2}{12} = \frac{7}{12}$	$\frac{2}{12} + \frac{1}{12} + \frac{1}{3} = \frac{7}{12}$
$\frac{9}{108} = \frac{1}{12}$	$\frac{2}{3} - \frac{1}{12} = \frac{7}{12}$	
[vii]	[viii]	[ix]

The stages of working this problem from [i] to [vi] on the counting board are clear and are probably displayed as follows:

I III	— II	⊥ ≡
II III	= IIII	
III IIII	= 丅	
[i]	[ii]	[iii]

≡ 丅	≡ 丅	I 丌
	⊥ II	
	⊥ I	
[iv]	[v]	[vi]

In stage [vii], we have assumed the fractions $\frac{18}{108}$ and $\frac{9}{108}$, though there is no explanation of how they are derived. We arrived at these fractions from the description of the averaging of fractions in *Jiu zhang suanshu* [Qian ed. 1963, p. 98]. The passage is translated below. (The lower case Roman numbers inserted in the passage refer to the stages of the above problem.)

> Multiply mutually each numerator and the other denominators [ii]; add [the products] and call the sum *ping shi* 平實 (lit. the average dividend) [iii]. Multiply the denominators to form the divisor (*fa*) [iv]. Multiply each [of the products] before their addition by the number of displayed [fractions] and call each value *lie shi* 列實 (lit. the displayed dividend) [v]; also, multiply the divisor by the number of displayed [fractions] [vi]. Subtract the *ping shi* from each [larger] *lie shi*. [The fractions formed from] the remainders [and the divisor] are reduced to give the required subtrahends [vii]. The sum of the subtrahends is added to the smaller [fraction] [ix]; assign the *ping shi* to the divisor (*yi fa ming ping shi* 以法命平實); each attains the average value.

The products 36, 72 and 81 shown in [v] are called *lie shi* 列實 (the displayed dividends). The sum, 63, in [iii], called *ping shi* 平實, which means "average dividend", is subtracted from each of the two larger *lie shi*, i.e., 81 and 72 to give 18 and 9 respectively. This explains how the fractions of [vii] are formed and then reduced. As shown in [viii], these fractions are the respective subtrahends of the two larger fractions given in the problem. Both subtractions result in the same difference, which is the average value of the three given fractions. When the two subtrahends are added and the sum is added to the smallest given fraction, the result is once again the average value of the three fractions as shown in [ix].

Stage [viii] involves the subtraction of fractions; the arrangement of rod numerals displayed here follows the description on the subtraction of fractions (see p. 84). In the same way, stage [ix] follows the procedure for the addition of fractions (p. 83).

A direct method for finding the average value of a set of fractions is stated in the last bit of the above passage: "assign the *ping shi* to the divisor". The divisor here is the product of the denominators of the fractions multiplied by the number of fractions (see [vi]). For example, suppose there are three fractions, $\frac{a}{x}, \frac{b}{y}, \frac{c}{z}$, the *ping shi* is $ayz + bxz + cxy$, the divisor is $3xyz$ and the average value is $\frac{ayz+bzx+cxy}{3xyz}$.

4.5   Other rules

The *Jiu zhang suanshu* gives the earliest systematic account of fractions in the world. The general rules governing fractions that we know today are stated there. Besides the above rules which Sun Zi had chosen for beginners, the book also discusses the multiplication and division of fractions, and the conversion of a mixed number into an improper fraction.

The rule for the multiplication of fractions (*cheng fen* 乘分) is stated as follows [Qian ed. 1963, p. 100]:

> Multiply the denominators (*mu* 母) to obtain the divisor (*fa*). Multiply the numerators (*zi* 子) to obtain the dividend (*shi*). Divide (*shi ru fa er yi*).

The rule for the multiplication of mixed numbers (such as $3\frac{1}{3} \times 5\frac{2}{5}$) is as follows [Qian ed. 1963, p. 101]:

> For each [mixed number], multiply the whole [number] by the denominator (*mu* 母) and add [the product] to the numerator (*zi* 子). Multiply [the numerators] to obtain the dividend (*shi*). Multiply the denominators (*mu*) to obtain the divisor (*fa*). Divide (*shi ru fa er yi*).

The division of fractions called *jing fen* 經分 is illustrated in the following two examples [Qian ed. 1963, pp. 98–99]:

> Now if $8\frac{1}{3}$ *qian* 錢 are divided among 7 persons, find how much each gets.
> Answer: Each person gets $1\frac{4}{21}$ *qian*.

> Next if $6\frac{1}{3}$ *qian* [plus] $\frac{3}{4}$ *qian* are divided among $3\frac{1}{3}$ persons, find how much each gets.
> Answer: Each person gets $2\frac{1}{8}$ *qian*.
> Method: Let the number of persons be the divisor (*fa*) and the amount of money be the dividend (*shi*). Divide (*shi ru fa er yi*). If [a number] has a fractional part, it has to be "generalised" (*tong* 通) [into an improper fraction].

In explaining the conversion of a mixed number into an improper fraction, Liu Hui made the following comment [Qian ed. 1963, p. 99]:

> Multiply the whole number [by the denominator] to disperse (*san* 散) [the whole number] into "accumulated parts" (*ji fen* 積分). Since these "accumulated parts" have been "generalised" (*tong* 通) mutually with the numerator of the fraction, they are thus joined to it.

In the reduction of a fraction, the method of successive subtractions enables the *deng shu*, which is the greatest common divisor of the numerator and denominator, to be found. This method is equivalent to the Euclidean algorithm of finding the greatest common divisor of two numbers. The solutions of Probs. 3, 4, 6–10 in Chapter 4 of *Jiu zhang suanshu* reveal that the ancient Chinese also knew how to find the least common multiple of a set of numbers. These problems involve the addition of fractions; the denominator of the sum of the fractions is stated and this is the least common multiple of their denominators. For example, in Prob. 10, the common denominator of the following sum

$$1 + \frac{1}{2} + \frac{1}{3} + \frac{1}{4} + \frac{1}{5} + \frac{1}{6} + \frac{1}{7} + \frac{1}{8} + \frac{1}{9} + \frac{1}{10} + \frac{1}{11}$$

is given as 27,720 which is the least common multiple of $\{1, 2, 3, 4, 5, 6, 7, 8, 9, 10, 11\}$.

4.6   The subject in general

The study of fractions is today an important part of arithmetic. In ancient China the same subject with an analogous set of rules formed a fundamental and significant part of mathematics. The Chinese developed their knowledge of fractions through the rods — the rapid expansion of the subject was due to the representation of the concept of a fraction in a simple elegant notation by means of rod numerals. This notation was evolved from the remainder of the arithmetical procedure of division, and it is the same as the notation we use today. History has shown that while the notation adopted by the Chinese had allowed for the thorough development of the subject, the notations assumed by other ancient civilizations, such as Babylonian, Egyptian, Greek and Roman, had impeded the expansion of some aspect or other of the subject. (See [Flegg ed. 1989, pp. 131–152; Karpinski 1925, pp. 121–125].)

The 5th century mathematician, Zhang Qiujian 張邱建, began his mathematical treatise, *Zhang Qiujian suanjing* 張邱建算經 (The mathematical classic of Zhang Qiujian), with fractions. His problems on fractions are of a more difficult nature than those in *Jiu zhang suanshu*, and the methods like those in *Sun Zi suanjing* provide us with useful information on how fractions were computed with counting rods (see [Ang 1969, pp. 13–20; Lam 1987, p. 386]). In the preface of his book Zhang made the following remarks on fractions [Qian ed. 1963, p. 329]: "In learning computation, one is not perturbed by the difficulties in the multiplication and division [of numerals], but by the difficulties in the manipulation of fractions". He was obviously referring to computation with rod numerals when he went on to say, "It is necessary [to know] the fundamentals concerning the orderly arrangements of fractions, and to be clear on the essential methods of manipulating with them".

It is difficult to say when the Chinese began to compute with fractions. What is certain is that their encounter with fractions was inextricably entangled with calendar making. Recognising that the synodic month and the tropical year were two incommensurable periods, the ancient Chinese had to devise methods to deal with them. This is evident in six ancient calendars produced between 246 BC to 207 BC [Chen 1955, p. 30]. For example, in the Zhuan Xu 顓頊 calendar used during the Qin dynasty (221–201 BC), a year of $12\frac{7}{19}$ months was taken to be $365\frac{1}{4}$ days, and the average length of a month was calculated as follows:

$$365\tfrac{1}{4} \div 12\tfrac{7}{19} = \tfrac{1461}{4} \div \tfrac{235}{19} = \tfrac{1461\times19}{4\times235} = \tfrac{27759}{940} = 29\tfrac{499}{940}.$$

Subsequently such astronomical calculations were systematically recorded in *Zhou bi suanjing* 周髀算經 — the following example of calculating the travel of the moon per year shows the magnitude of the treatment of fractions [Qian ed. 1963, p. 68]:

1 month = $29\tfrac{499}{940}$ days

12 months = $29\tfrac{499}{940} \times 12 = 354\tfrac{348}{940} = \tfrac{333108}{940}$ days.

Daily lunar motion = $13\tfrac{7}{19} = \tfrac{254}{19}$ degrees.

12 monthly lunar motion = $\tfrac{254}{19} \times \tfrac{333108}{940} = \tfrac{84609432}{17860} = 4737\tfrac{6612}{17860}$ degrees.

The treatment of such fractions was necessitated by the rise and development of astronomy in ancient China. The use of fractions could therefore be traced back many centuries before any mathematical text was written.

On Extracting Roots of Numbers

5.1 The background

Like the basic operations of arithmetic and the rules of the common fraction, it is not known when the ancient Chinese first knew how to find the square root of a number. This method and that of finding the cube root are recorded in *Jiu zhang suanshu* [Qian ed. 1963, pp. 150 & 153]. Because of the terseness and obscurity of the text, it is difficult to infer an unambiguous step by step explanation of how roots were extracted through the medium of counting rods. Most scholars such as Wang Ling & Needham [1955, pp. 350–356], Qian Baocong [1964, pp. 47–48], Li Yan 李儼 & Du Shiran 杜石然 [1987, pp. 50–51] and Bai Shangshu 白尚恕 [1983, pp. 104–108] have the same interpretation for the general method of finding the square root, and this has been accepted by others such as Li Di 李迪 [1984, pp. 69–77] and Shen Kangshen 沈康身 [1986, p. 179].[1] Their explanation of finding the integral part of the square root is quite similar to that described in *Sun Zi suanjing*.

Sun Zi explained the method by illustrating it in two examples (Ch. 2, Probs. 19 & 20). In contrast with the difficult style of *Jiu zhang suanshu*, the description of the root extraction method in *Sun Zi suanjing* is simpler

---

[1] Xu Xintong 許鑫銅 [1987] has a different interpretation.

and more lucid. An analysis of the latter will therefore provide an easier access to the understanding of the origins of an important subject in the history of mathematics in China.

5.2   The method in *Sun Zi suanjing*

Ch. 2, Prob. 19 states:

> Now there is an area of 234,567 *bu* 步, find [one side of] the square. Answer: $484\frac{311}{968}$ *bu*.

We give below a step by step exegesis of the method. In each step, an interpretation of how the rod numerals would be displayed on the counting board is shown, and sometimes further explanation is added.

Step 1

> Put down the area, 234,567 *bu*, as *shi* 實 (lit. dividend).

$$= ||| \equiv |||| \perp \top \qquad shi \qquad 234567$$

Step 2

> Next use one rod as *xia fa* 下法 (lit. lower divisor). [Move this rod from below] the units in *bu* [of the *shi*] overpassing one place (*chao yi wei* 超一位) to reach the hundreds and stop (*zhi bai er zhi* 至百而止).

$$= ||| \equiv |||| \perp \top \qquad shi \qquad 234567$$
$$| \qquad xia\,fa \qquad 1$$

$$= ||| \equiv |||| \perp \top \qquad shi \qquad 234567$$
$$| \qquad xia\,fa \qquad 1$$

Explanation: For this particular problem which has a three-figured root, there is an error in the wording of the last sentence. The word "hundreds" should be replaced by "ten thousands" and the sentence should read somewhat like this: "[Move this rod from below] the units in *bu* [of the *shi*] overpassing one place [at a time] to reach the ten thousands and stop." The computation with rod numerals would then appear as follows:

=|||≡||||⊥𝍔   *shi*   234567

| *xia fa*   1

This correction is justified because in Steps 7 & 12, we are asked to move this rod back towards the right two places at a time so that ultimately it is once again below the units of the *shi*. The same error was repeated by Zhang Qiujian 張邱建 in all his problems which involved the extraction of a three-figured square root [Qian ed. 1963, pp. 369–371]. Yang Hui 楊輝 gave the correct version of this step when he stated [*Yongle dadian* Ch. 16,344, p. 7b], "Put down the area as *shi*. Separately, put down one rod as *xia fa*. [Place] this rod below the *shi*, and [move] it from the last place [i.e. the units] [of the *shi*] by overpassing successively one place [at a time]. Stop when [the rod] reaches the beginning of the *shi*." Yang Hui went on to give the following explanatory note: "When [the *xia fa*] is below the units [of the *shi*], this determines the units [of the roots]; when it is below the hundreds, this determines the tens [of the root]; when it is below the ten thousands, this determines the hundreds [of the root]; when it is below the thousand thousands, this determines the thousands [of the root]."

Step 3

Put down 400 as *shang* 商 (lit. quotient) above the *shi*.

|||| *shang*   4

=|||≡||||⊥𝍔   *shi*   234567

| *xia fa*   1

Explanation: The determination of the digit, 4, for the hundreds of the root is through trial and error. It is the largest possible digit which, after following the procedure of Steps 4 & 5, leaves a non-negative numeral for the *shi* in Step 5. This digit is placed above the hundreds of the *shi* and its position conforms with Yang Hui's explanation since the *xia fa* is below the ten thousands of the *shi*.

Step 4

Next put down 40,000 below the *shi* and above the *xia fa* and call it *fang fa* 方法 (lit. square divisor).

IIII	*shang*	4
= III ☰ IIIII ⊥ 𝕋	*shi*	234567
IIII	*fang fa*	4
I	*xia fa*	1

Explanation: Having obtained the digit for the hundreds of the root, the same digit is also placed in the row immediately below the *shi* and in the same column as that of the single rod of the *xia fa*. In this position, it represents 40,000 for the *fang fa*.

Step 5

Assign (*ming* 命) the *shang*, 400, in the top position to it [in order] to subtract from the *shi*.

IIII	*shang*	4
𝕋 ☰ IIIII ⊥ 𝕋	*shi*	74567
IIII	*fang fa*	4
I	*xia fa*	1

Explanation: In this context, "assign" (*ming*) means to use the digit of the *shang* to multiply the digit of the *fang fa*. The place value of the product corresponds with the digit of the *fang fa* so that the product, 16, is immediately below the "23" of the *shi*. The subtraction of these two numerals leaves "7" in place of "23", so the *shi* is now 74,567. (For the method of subtraction, see pp. 72–73.)

Step 6

After subtraction double the *fang fa*.

ⅠⅠⅠⅠ	*shang*	4
𝍒 ☰ ⅠⅠⅠⅠ ⊥ 𝍒	*shi*	74567
𝍓	*fang fa*	8
Ⅰ	*xia fa*	1

Step 7

Shift (*tui* 退) the *fang fa* [to the right] by one place and the *xia fa* by two places.

ⅠⅠⅠⅠ	*shang*	4
𝍒 ☰ ⅠⅠⅠⅠ ⊥ 𝍒	*shi*	74567
⏊	*fang fa*	8
Ⅰ	*xia fa*	1

Explanation: The single rod of the *xia fa* in the hundreds' place indicates the determination of the tens' digit for the root.

Step 8

Put down 80 in the top position as *shang* next to the previous *shang*.

⦀⫶	shang	48
𝕋 ☰ ⦀⦀ ⊥ 𝕋	shi	74567
⊥	fang fa	8
		1
Ɩ	xia fa	1

Explanation: The determination of the digit 8 for the tens of the root is through trial and error. It is the largest possible digit which, after following the procedure of Steps 9 & 10, leaves a non-negative numeral for the *shi* in Step 10. This digit is placed above the tens of the *shi*.

Step 9

Also put down 800 below the *fang fa* and above the *xia fa* and call it the *lian fa* 廉法 (lit. side divisor).

⦀⫶	shang	48
𝕋 ☰ ⦀⦀ ⊥ 𝕋	shi	74567
⊥	fang fa	8
𝕋	lian fa	8
Ɩ	xia fa	1

Explanation: Having obtained the digit for the tens of the root, the same digit is also placed in the row immediately below the *fang fa* and in the same column as that of the single rod of the *xia fa*.

Step 10

Assign (*ming*) the *shang*, 80, in the top position to each of the *fang*
and *lian* [in order] to subtract from the *shi*.

IIII ⊥	*shang*	48
I IIIII ⊥ 丅	*shi*	10567
⊥	*fang fa*	8
丅	*lian fa*	8
I	*xia fa*	1

IIII ⊥	*shang*	48
☰ I ⊥ 丅	*shi*	4167
⊥	*fang fa*	8
丅	*lian fa*	8
I	*xia fa*	1

Explanation: The digit for the tens of the *shang* multiplies the digit of the
fang fa. The product, 64, is immediately below the "74" of the *shi*. The
subtraction of the two numerals leaves "10" in place of "74", so the *shi* is
now 10,567. Next the digit for the tens of the *shang* multiplies the digit of
the *lian fa*. The product, 64, is immediately below the "105" of the *shi*. The
subtraction of the two numerals leaves "41" in place of "105", so the *shi* is
now 4,167.

Step 11

After subtraction, double the *lian fa* and join it to the *fang fa* above.

IIII ⊥	shang	48
☰ I ⊥ Ⱦ	shi	4167
⊥ T	fang fa	96
I	xia fa	1

Step 12

Shift the *fang fa* [to the right] by one place and the *xia fa* by two places.

IIII ⊥	shang	48
☰ I ⊥ Ⱦ	shi	4167
Ⲧ ⊥	fang fa	96
I	xia fa	1

Explanation: The single rod of the *xia fa* in the units' place indicates the determination of the units' digit for the root.

Step 13

Put down 4 in the top position as *shang* next to the previous [*shang*].

IIII ⊥ IIII	shang	484
☰ I ⊥ Ⱦ	shi	4167
Ⲧ ⊥	fang fa	96
I	xia fa	1

Explanation: The determination of the digit, 4, for the units of the root is through trial and error. It is the largest possible digit which, after following the procedure of Steps 14 & 15, leaves a non-negative numeral for the *shi* in Step 15.

Step 14

Also put down 4 below the *fang fa* and above the *xia fa* and call it *yu fa* 隅法 (lit. corner divisor).

||||⫢ ||||	shang	484
≡	⊥ ╥	shi	4167
╥⊥	fang fa	96
||||	yu fa	4
		xia fa	1

Explanation: Having obtained the digit for the units of the root, the same digit is also placed in the row immediately below the *fang fa* and in the same column as that of the single rod of the *xia fa*.

Step 15

Assign (*ming*) the *shang*, 4, in the top position to each of the *fang*, *lian* and *yu* [in order] to subtract from the *shi*.

|||| ⫢ ||||	shang	484
|||||⊥ ╥	shi	567
╥⊥	fang fa	96
||||	yu fa	4
		xia fa	1

IIII ⊥ IIII	*shang*	484
III = TT	*shi*	327
TTT ⊥	*fang fa*	96
IIII	*yu fa*	4
I	*xia fa*	1

IIII ⊥ IIII	*shang*	484
III — I	*shi*	311
TTT ⊥	*fang fa*	96
IIII	*yu fa*	4
I	*xia fa*	1

Explanation: The digit for the units of the *shang* multiplies the digits of the *fang fa*, which includes the *lian fa* (see Step 11). Following the method of multiplication (see Sect. 3.3.1), 4 first multiplies 9 to give 36 and this is subtracted from the "41" above to leave 567 for the *shi*. Next 4 multiplies 6 to give 24 and this is subtracted from the "56" above to leave 327 for the *shi*. The digit for the units of the *shang* also multiplies the digit of the *yu fa*. The product, 16, is immediately below the "27" of the *shi*. The subtraction of the two numerals leaves "11" in place of "27", so the *shi* is now 311.

Step 16

After subtraction, double the *yu fa* and join this to the *fang fa*.

IIII ⊥ IIII	*shang*	484
III — I	*shi*	311
TTT ⊥ TT	*fang fa*	968
I	*xia fa*	1

Explanation: This step is in line with the procedure of Steps 6 and 11.

Step 17

The *shang* in the top position has 484 and the *fa* in the lower position has 968, while the remainder is 311. Hence [one side of] the square is $484\frac{311}{968}$ *bu*.

IIII ⩦ IIII	*shang*	484
III — I	*shi*	311
ℿ ⊥ ℿ	*fa*	968

Explanation: The *xia fa* is discarded and the *fang fa* is simply called *fa* (divisor).

We have already drawn attention to the similarities between the Chinese and Arabic methods of computation in the basic operations of arithmetic. When a comparison is drawn between Sun Zi's method of extracting the square root and the earliest Arabic method, their resemblance is once again strikingly close. (See [Saidan 1978, pp. 76–79; Levey & Petruck 1965, pp. 64–68].)

5.3   An analysis of the method

5.3.1   A comparison with the method of division

The method is a general one which can be applied mechanically to derive the square root of any number. There are only two problems on root extraction in *Sun Zi suanjing* and both are on finding the square root of a number of six digits. The second problem (Ch. 2, Prob. 20) involves the square root of 420,000, which is given as $648\frac{96}{1296}$, and the instructions for solving are identical with those given in the first problem.

There is a remarkable similarity between the procedures for division (Sect. 3.4.1) and the extraction of the square root; one can even say that the latter is a complicated extension of the former. We now point out the main parallelisms.

The dividend is called *shi* and the number whose root is to be extracted is also called *shi*. This is the first numeral to be put on the counting board, and its digits set the values of the places for the other digits on the board.

In division, the divisor termed *fa* is moved from right to left such that its first digit from the left is placed either below the first digit (from the left) of the dividend or below the second digit of the dividend, depending on the value of the dividend. The extraction of a root requires more than one operator, and the number of operators depends on how large the *shi* is. All the operators are called *fa* and they are differentiated from one another by the different names prefixed to *fa*. In the above example, the operators are *xia fa, fang fa, lian fa* and *yu fa*. The operation begins with using a unit rod called *xia fa*, which means "the lower divisor". This rod is first placed below the units of the *shi* and is moved from right to left by crossing over one place at a time. From the units, it jumps to the hundreds, then to the ten thousands, and then to the thousand thousands, and so on, depending on the extent of the *shi*. It stops below the second or first digit (from the left) of the *shi*.

In division, when the units of the divisor are below the tens of the dividend, this position determines the tens of the quotient; when they are below the hundreds of the dividend, this determines the hundreds of the quotient, and so on (pp. 63–64). In square root extraction, the *xia fa* jumps by steps of two places; when it is below the units of the *shi*, this position determines the units of the root; when it is below the hundreds of the *shi*, this determines the tens of the root; when it is below the ten thousands of the *shi*, this determines the hundreds of the root; and so forth.

In division, each digit of the quotient is determined through a trial and error process. After the digit is obtained, it multiplies the divisor (*fa*), and the product, whose place value is determined by the position of the divisor (*fa*), is subtracted from the corresponding digits of the dividend (*shi*) above it. This process of subtracting or "removing" from the dividend (*shi*) is called *chu* 除. In the case of root extraction, though the procedure for finding each digit of the root is more complicated, the process is also through trial and error. After the digit is obtained, it multiplies certain operators called *fa*. The place values of the products are determined by the positions of the operators, and the numerals are subtracted from the corresponding digits of the *shi* above them. Once again, this process of subtracting or "removing" from the *shi* is called *chu*.

The process of subtracting from the *shi* stage by stage results eventually in either a blank or a remainder in the *shi*. If there is no remainder in the *shi* in division then the quotient is an integer, and if there is a remainder the quotient has a fraction. Likewise, if there is no remainder in root extraction the root is integral, and if there is a remainder the root can be expressed with either an appropriate or an approximate fraction.

5.3.2 An approximation for an irrational root

In Sun Zi's two examples, both roots are irrational and the given method illustrates the expression of the root in an approximate rational form.

The method in *Jiu zhang suanshu* explains only the computation of the integral portion of the root; should the root of an example be irrational, the computation of the remaining portion of the *shi* is described in four characters only: *yi mian ming zhi* 以面命之 (lit. use a side to assign). Wang Ling & Needham [1955, pp. 356, 376] drew attention to this phrase, which according to them indicated that the Han mathematicians were able to continue their operations even when the root was non-integral. In another piece of work, Wang Ling [1956, p. 255] suggested that this phrase meant the extension of the root extraction to a common fraction of the form $r/(2a + b)$, where r is the final remainder of the *shi*, a the integral portion of the root and b the first digit of its decimal portion. Xu Chunfang 許莼舫 [1965, p. 29] gave another interpretation of the phrase by saying that if $A = a^2 + r$, where $r > o$, then $\sqrt{A} \approx a + r/a$. Xu Chunfang however admitted that this approximate value was far from satisfactory. As there are no examples in *Jiu zhang suanshu* of how an irrational root is expressed, the true meaning of the phrase *yi mian ming zhi* remains ambiguous. From Liu Hui's 劉徽 commentary of this phrase, Bai Shangshu 白尚恕 [1983, pp. 108–109] stated that Liu Hui was aware of two approximations, namely, $a + r/2a$ and $a + r/(2a + 1)$, and the fact that $a + r/(2a + 1) < \sqrt{A} < a + r/2a$.

Since Sun Zi's example gives the square root of 234,567 as $484\frac{311}{968}$, this depicts the approximation as $a + r/2a$. This formula was known to Heron of Alexandria [Cajori 1893, pp. 43–44]; C. N. Srinivasiengar [1967, pp. 14] claimed that the formula was known in ancient India. What is significant about Sun Zi's method is that it is probably the earliest that showed step by step how this form of approximation was derived. His method also gives the clue for the derivation of the other common form of approximation, namely, $a + r/(2a + 1)$, which was used by mathematicians such as Zhang Qiujian [Qian ed. 1963, pp. 370–371]. The extra unit in the denominator of the fraction indicated that the unit rod of the *xia fa* was included (see p. 102, Step 16).[2]

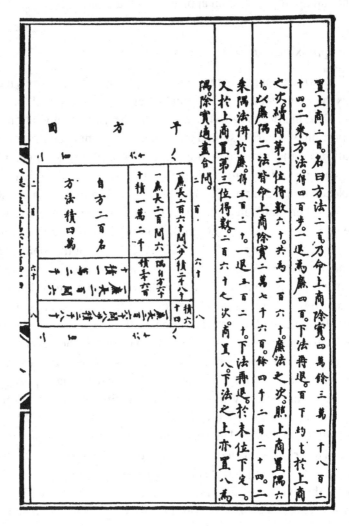

Fig.5.1   The oldest surviving diagram on the derivation of the square root extraction method preserved in *Yongle dadian* [Ch. 16,344, p. 8a].

---

[2] This assumption is reasonable since 2*a* is the sum of the numerals representing the operators *fang fa* and *yu fa* (Step 16), and the *xia fa*, which is the remaining operator of the root extraction procedure, is represented by a unit rod.

5.3.3 The geometrical derivation and its arithmetization

It is generally believed that the concept of the square root method was based on geometrical considerations. This is supported by the meanings of some technical terms, which were used in *Jiu zhang suanshu* [Wang & Needham 1955, pp. 394–395]. Furthermore Liu Hui's commentary of the passage demonstrates the use of coloured diagrams [Qian ed. 1963, pp. 150–151]. In Sun Zi's description, he not only continued to use terms like *fang fa* (divisor of the square) and *chu* (to remove), but also employed terms such as *lian fa* (divisor of the side area) and *yu fa* (divisor of the corner area); all these have geometrical implications.

The oldest surviving diagram on the derivation of the square root extraction method is found in Yang Hui's *Xiangjie jiu zhang suanfa* 詳解九章算法 (A detailed analysis of the mathematical methods in the "Nine Chapters") of 1261; it is preserved in *Yongle dadian* 永樂大典 [Ch. 16,344, p. 8a] (Fig. 5.1). The diagram illustrates a problem from *Jiu zhang suanshu* (Ch. 4, Prob. 14), which is to find the square root of 71,824. In addition to the diagram, Yang Hui explained the method and showed the different stages on the counting board (Fig. 5.2). (See [Lam 1969b, pp. 93–97].)

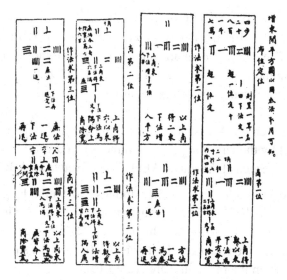

Fig. 5.2 Yang Hui's method on the counting board taken from *Yongle dadian* [Ch. 16,344, pp. 8b–9a].

Based on Yang Hui's diagram, we construct a similar diagram for finding the square root of 234,256 (Fig. 5.3). This number is derived from the number 234,567 of Sun Zi's example minus the remainder, 311 (see Step 15). With the help of this diagram, we now describe briefly the correlation of the geometrical concept and the arithmetization of the concept on the counting board.

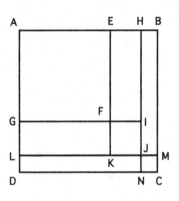

Fig. 5.3

The area of the square ABCD is 234,256. AE represents the length in hundreds, i.e., 400; EH the length in tens, i.e., 80; and HB the length in units, i.e., 4. The derivation entails three stages: the removal of (i) square AEFG, (ii) gnomon EHJLGF and (iii) gnomon HBCDLJ.

We shall now discuss each stage in turn.

(i)   When the length AE is obtained (Step 3), a similar length AG is identified by the term *fang fa*, which means "a divisor of the square" (Step 4). The area of the square AEFG is removed from ABCD (Step 5).

(ii) When AG is doubled (Step 6), this implies two similar lengths, EF and GF. When the length EH is obtained (Step 8), a similar length GL is identified by the term *lian fa*, which means "a divisor of the side area" (Step 9). The two rectangles EHIF and GFKL and the square FIJK are removed from the remaining area (Step 10).

(iii) When GL is doubled, this implies two similar lengths, IJ and KJ. When they are joined to EF (=HI) and GF (=LK) respectively, we have HJ and LJ respectively (Step 11). When the length HB is obtained (Step 13), a similar length LD is identified by the term *yu fa*, which means "a divisor of the corner area" (Step 14). The gnomon HBCDLJ is the last residue of the area to be removed (Step 15).

The computation of the area of a rectangle or a square was among the earliest topics on multiplication in ancient China. The first chapter of *Jiu zhang suanshu* began with four problems on finding the area of a rectangle followed by six similar problems (Probs. 19–24) which involved fractions. The first eleven problems of Chapter 4 involved the solution of an unknown side of a rectangle from a known area and side, and this was followed by five problems on finding the side of a square. All these problems involved a geometrical derivation, but the computation was performed through rod numerals.

Rectangles and squares are basic geometric concepts, and the derivation of the side of a square through geometric means could occur independently in more than one civilization. Theon of Alexandria (c. 390) employed a diagram similar to Yang Hui's, when he explained Ptolemy's method of extracting square roots with sexagesimal fractions [Gow 1923, pp. 55–57; Heath 1921, vol. 1, pp. 60–63].

What was unique in the Chinese invention was the method of computation through rod numerals. The ancient Chinese were able to transcribe their geometric understanding of the different stages of the square root diagram into parallel stages of computation on the counting board. In so doing they developed a general method for computing the square root of any number. This process of arithmetization with rod numerals was obviously based on their knowledge of the fundamental operations of arithmetic — we have already noted the close relationship between the division process and the square root method. The invention of a procedural method for division through the rod numeral system would therefore be an important foundation for the development of a general method of finding the square root of any number.

All the arithmetic methods on the counting board showed a common characteristic — the rows in which the numerals were set also stood for the technical terms of what the numerals represented, such as *shi, fa, shang*, etc. This form of representation would later evolve into notational expressions for equations as well. The complicated way of positioning the digits of each numeral relative to those of other numerals on the board played a necessary role in the successful operation of each method.

5.4   Significance of the method

The arithmetization through rod numerals of a diagrammatic concept led to the discovery of a general method which could mechanically extract the square root of a number. The method was, however, not only used as an algorithm — it provided the basis for the development of further methods. These methods include the extraction of the cube root of a number, the solutions of quadratic, cubic and polynomial equations of higher order. At the same time the notations necessary for conveying the meaning of these equations were also inextricably expressed in the positions occupied by the rod numerals on the counting board. It is beyond the scope of this book to explain in detail the development of polynomial equations, especially when there is no discussion of them in *Sun Zi suanjing*. We shall, instead, mention briefly a few cornerstones in the development.

The rows in which the numerals were set played the role of notations representing the concepts of their respective technical terms. In the different stages of the square root method, the three terms *shi, fang fa* and *xia fa* are associated with the formation of what we now call a quadratic equation of the form $x^2 + bx = c$, where b and c are positive. For example, Step 7 of Sect. 5.2 is associated with $y^2 + 800y = 74,567$ and Step 12 with $z^2 + 960z = 4,167$. Their method of solution is a part of the square root method. *Jiu zhang suanshu* has a problem (Ch. 9, Prob. 20) involving an equation of this form. The Han mathematicians were able to solve this type of quadratic equation based on the procedure of extracting the square root of a number.

The method of extracting the cube root of a number is similar to the square root method; in fact, it can be considered as an extension (see [Wang & Needham 1955, pp. 356–365]). Just as a part of the square root method can be used to solve a quadratic equation, so a section of the cube root method can be applied to solve a transformed cubic equation of the form

$x^3 + ax^2 + bx = c$, where a, b and c are positive. This type of equation is found in the 7th century book *Ji gu suanjing* 緝古算經 (Apprehending ancient mathematics) by Wang Xiaotong 王孝通.

Yang Hui quoted the 11th century mathematician Liu Yi 劉益 when he discussed the geometric derivation and the arithmetic computation of the various types of quadratic equations [Lam 1977, p. 112–125, 251–279; Qian et al. 1966, pp. 44–47]. While it was not possible to abstract a common geometric method from the variety of diagrams representing different types of quadratic equations, it was revealed that when the methods were transcribed into operations on the counting board, there were certain patterns and similarities in the operations which made possible the evolution of a general method on the counting board.

Yang Hui mentioned another 11th century mathematician, Jia Xian 賈憲, who, in addition to the original method of extracting the square and cube roots, drew attention to a revised method called *zeng cheng fang fa* 增乘方法 (method of adding and multiplying) [*Yongle dadian* Ch. 16,344, pp. 6b, 16a–17b; Qian et al. 1966, pp. 37–40]. The latter was an algorithmic method that could be extended to extract roots of higher degree, and this subsequently provided the breakthrough in the solution of any numerical polynomial equation.

By the 13th century, mathematicians were familiar with the method of solving polynomial equations [Lam 1982]. This was complemented by their ability to express the intricate concept of a polynomial equation in a concise notation on the counting board. This notation was an extended form of the notation for a quadratic equation discussed earlier. For instance, $a_0x^n + a_1x^{n-1} + \ldots + a_{n-1}x + a_n = 0$ was represented in rod numerals in the following manner:

$$
\begin{array}{c}
a_n \\
a_{n-1} \\
\cdot \\
\cdot \\
\cdot \\
\cdot \\
a_1 \\
a_0
\end{array}
$$

where each a_i occupied a definite row in the column.

The development of the root extraction method which culminated in the concept and solution of equations took place over a period of more than a thousand years. It is one of the finest illustrations of the potential of the rod numeral system that allowed for the expression and generation of mathematical ideas.

5.5   Conclusion

It was quite natural that the idea of finding a side of a square from its area should originate from diagrams. The Chinese went a step further to arithmetize the geometrical method through the use of rod numerals. In so doing, a general method was evolved which could extract the square root of any number; sometimes the root was irrational, and this was then expressed with an approximate fraction.

Just as the development of the square root method was based on the knowledge of division, so the extraction of the cube root was based on that of the square root. The extractions of higher order roots were subsequently invented together with the methods of solving numerical polynomial equations. The intricate concept of a polynomial equation was notated through the rod numeral system in a concise and simple form, and this undoubtedly facilitated its solution.

chapter 6

Tables of Measures

6.1 Introduction

We digress here to discuss tables of measures since most mathematical problems are involved with some form of measures or other. The *Sun Zi suanjing* begins with tables for three systems of measures: length, weight and capacity (pp. 191–192). They were formulated primarily on the well-known measures standardized by Emperor Qin Shihuang in the third century BC. The measures for length, weight and capacity had only five principal units each. The units for length were: *fen* 分, *cun* 寸, *chi* 尺, *zhang* 丈 and *yin* 引; those for weight were: *zhu* 銖, *liang* 兩, *jin* 斤, *jun* 鈞 and *dan* 石; and those for capacity were: *yue* 龠, *ge* 合, *sheng* 升, *dou* 斗 and *hu* 斛 [*Han shu* Ch. 21A, pp. 966–969]. Unlike the measures of length and capacity which were decimal in nature, those of weight were quite varied.

The lowest unit for length in *Sun Zi suanjing* was fixed as *hu* 忽 which, according to Sun Zi, was the diameter of a freshly spun silk thread. Three other units were added between *hu* and *fen* 分. These were: *si* 絲, *hao* 毫 and *li* 釐. In place of *si*, *Sui shu* 隋書 [Ch. 16, p. 402] has a different name: *miao* 秒.

Sun Zi used the size of a grain of millet, that is, *su* 粟, as the lowest unit in the capacity measures. Since the size of a grain of millet was too minute for any practical purpose, he allocated six of these grains as the standard measure for *gui* 圭. This was followed by *cuo* 撮 and *chao* 抄; he then renamed *yue* 龠 as *shao* 勺, and left the names of the other units unchanged.

Another variety of millet, the glutinous grain called *shu* 黍, was used as the lowest unit of weight. The weight of ten of these grains constituted the next lowest unit, *lei* 絫. These two units were then joined to the lowest unit, *zhu* 銖, of the original table.

We now discuss each type of measure in turn.

6.2   Measures of length

The first part of the linear measure as stated by Sun Zi is entirely in the decimal system. It is as follows:

$$
\begin{array}{rcl}
10 \ hu \ 忽 & = & 1 \ si \ 絲 \\
10 \ si & = & 1 \ hao \ 毫 \\
10 \ hao & = & 1 \ li \ 釐 \\
10 \ li & = & 1 \ fen \ 分 \\
10 \ fen & = & 1 \ cun \ 寸 \\
10 \ cun & = & 1 \ chi \ 尺 \\
10 \ chi & = & 1 \ zhang \ 丈 \\
10 \ zhang & = & 1 \ yin \ 引.
\end{array}
$$

Chi 尺 was adopted in China as the basic unit of length throughout successive centuries; however its length varied from time to time. In his investigation of some twenty different kinds of *chi* from the Han dynasty to the turn of the 20th century, Wang Guowei 王國維 [1928] noted a tendency for the *chi* to increase in length, though the rate of increase had not always been uniform. He found the increase was greatest during the three hundred years that separated the Western Jin from the later Wei dynasty: 265 to 554 AD. From the tabulated values of *chi* in terms of centimetres, as given by Wu Chengluo 吳承洛 [1937, pp. 54 & 90], the length of one *chi* during the Western Jin period was 24.12 cm and at the end of the Wei period was 29.97 cm. The length of a *chi* during Sun Zi's time would probably be about 29.51 cm.

The measures smaller than *cun*, namely, *fen*, *li*, *hao*, *si* and *hu*, have a special significance; they were meant to express extremely small units of length beginning with one *hu*, which Sun Zi described as having the thickness of "a strand of silk vomited by a silkworm". As there was no mechanical device that could give the measure of one *hu*, the terms were just names for the decimal fractional parts of one *cun*. In fact, *fen* meant "a fractional part" or "a tenth", and it was also the technical term for a fraction. The names most probably arose from the necessity of computing a length as accurately as possible, as in the case of finding an approximation for an irrational value. For instance, the use of these terms can be found in Liu Hui's computation of π (see [Lam & Ang 1986, pp. 328, 336–337]).[1]

When a length such as 1 *cun* 2 *fen* 3 *li* 4 *hao* 5 *si* was transcribed on to the counting board, its concise rod numeral notation in the form

$$| = ||| \equiv |||||$$

expressed the metrological concept as well. The computer knew that the position occupied by the digit 1 indicated the units place in *cun*, while the next place to the right was called *fen*, and the successive places after it were named *li*, *hao* and *si* respectively.

When the concept of 1 *cun* 2 *fen* 3 *li* 4 *hao* 5 *si* is expressed in terms of our present decimal fraction notation, this appears as

$$1.2345 \; cun.$$

The two notations are strikingly similar. The written form has a dot or a decimal point to separate the integral part of a number from its fractional part.

In the rod numeral notation, *fen*, *li*, *hao*, *si* and *hu* were the names used by the computer to describe the first, second, third, fourth and fifth decimal places respectively of a numerical length in *cun*. The fundamental arithmetic operations and the extraction of roots were performed on such numerals in the same manner as they were performed on numerals representing integers. (See [Lam & Ang 1986, pp. 336–337].)

---

[1] Liu Hui called numbers of such measures *wei shu* 微數 meaning "minute numbers", which reflects an abstract concept of the decimal fraction. Instead of the name *si*, Liu Hui used the term *miao* 秒.

The other measures of length listed by Sun Zi are non-decimal and involve large quantities. They are

50 *chi* 尺	=	1 *duan* 端
40 *chi*	=	1 *pi* 疋
6 *chi*	=	1 *bu* 步
240 *bu*	=	1 *mu* 畝
300 *bu*	=	1 *li* 里.

Cun, chi, zhang and *bu* are very commonly used (Ch. 2, Probs. 9–13, 15–20, 22, 27 & Ch. 3, Probs. 3, 18, 22, 25). Measurements involving longer lengths such as that of a canal (Ch. 2, Prob. 23 & Ch. 3, Prob. 32) and the distance between Chang' an and Luoyang (Ch. 3, Prob. 33) employ *li* as well. The measure for cloth *pi* 疋 has a length of 40 *chi*; this is used in Ch. 3, Prob. 21 in the discussion on the cost of brocade. On the subject of thin silk in Ch. 2, Prob. 28, Ch. 3, Probs. 8 & 16, where length is not mentioned, another character *pi* 匹, which generally means "roll", is used.

The same names for lengths are used for areas and volumes; for example, the area of a square of side 2 *bu* is called 4 *bu* and the volume of a cube of side 2 *bu* is called 8 *bu*. *Mu* is a measure for the area of a field equivalent to an area of 240 *bu*. One *qing* 頃 is 100 *mu*, which is used in Ch. 2, Probs. 14 & 21, but is not listed in the given tables.

When measures are not mixed, the measures for areas and volumes are straightforward, so that the area of a square of side 2 *chi* is 4 *chi* and the volume of a cube of side 2 *chi* is 8 *chi*. A study of some of the problems reveals that when there are mixed measures in *zhang, chi* and *cun*, a decimal system is adopted in terms of *chi*. For instance, in Ch. 2, Prob. 23, when 1 *zhang* 3 *cun* is multiplied by 1 *zhang* 8 *chi*, the result is given as "185 *chi* 4 *cun*" (which in our context means 185.4 sq. *chi* since the product is 185 sq. *chi* 40 sq. *cun*). Proceeding from here, we are asked to multiply this by a length of 52,824 *chi* to obtain a volume of "9,793,569 *chi* 6 *cun*".

In Ch. 2, Probs. 10–12 & Ch. 3, Prob. 3, the volume capacity to hold 1 *hu* 斛 is stated as 1 *chi* 6 *cun* 2 *fen*; our interpretation of this figure should be 1.62 cu. *chi*. In Ch. 3, Prob. 25, the measure of the product of 1 *zhang* 5 *chi* and 1 *chi* 5 *cun* is once again treated linearly as 2 *zhang* 2 *chi* 5 *cun* — we are asked to raise this figure to tens (*shang shi zhi* 上十之) to obtain 22 *zhang* 5 *chi* and to divide by 5 *cun* to obtain the answer, which is 4 *zhang* 5 *chi*.

6.3   Measures of weight

The weight measures listed at the beginning of *Sun Zi suanjing* are as follows:

$$
\begin{aligned}
10\ shu\ \text{黍} &= 1\ lei\ \text{絫} \\
10\ lei &= 1\ zhu\ \text{銖} \\
24\ zhu &= 1\ liang\ \text{兩} \\
16\ liang &= 1\ jin\ \text{斤} \\
30\ jin &= 1\ jun\ \text{鈞} \\
4\ jun &= 1\ dan\ \text{石}.
\end{aligned}
$$

Of these measures, only *liang* and *jin* are used in the problems (Ch. 3, Probs. 9, 10, 14 & 20).

6.4   Measures of capacity

The measures of capacity listed in *Sun Zi suanjing* are as follows:

$$
\begin{aligned}
6\ su\ \text{粟} &= 1\ gui\ \text{圭} \\
10\ gui &= 1\ cuo\ \text{撮} \\
10\ cuo &= 1\ chao\ \text{抄} \\
10\ chao &= 1\ shao\ \text{勺} \\
10\ shao &= 1\ ge\ \text{合} \\
10\ ge &= 1\ sheng\ \text{升} \\
10\ sheng &= 1\ dou\ \text{斗} \\
10\ dou &= 1\ hu\ \text{斛}.
\end{aligned}
$$

Qian [ed. 1963, p. 275] noted that the values of *cuo* and *chao* were interchanged in *Sui shu* so that

$$
\begin{aligned}
10\ gui &= 1\ chao \\
10\ chao &= 1\ cuo \\
10\ cuo &= 1\ shao.
\end{aligned}
$$

Except for 6 *su* = 1 *gui*, the rest of the measures are in decimal form. In Ch. 2, Probs. 10–12 & Ch. 3, Prob. 3, we are told that 1 *hu* has a volume capacity of 1 *chi* 6 *cun* 2 *fen*; the interpretation of this measure has already been discussed. A large number of problems involving capacity measures are on exchange (Ch. 2, Probs. 5–8 & Ch. 3, Probs. 11 & 16), and the

others are on miscellaneous topics (Ch. 3, Probs. 1, 6, 12, 13, 19 & 23).
Measures smaller than *shao* are not used in the book.

6.5   A common set of decimal fractional units

It is not known how early the set of names, *fen* 分, *li* 釐, *hao* 毫, *si* 絲 and
hu 忽, were employed to express decimal fractional parts of measures other
than *cun*. In *Sun Zi suanjing* Ch. 3, Prob. 2, the rate of conscription of the
number of men to one soldier is given as 37 men 5 *fen*. This expression is
interesting not only because this is the only instance in the book where *fen*
is attached to a measure other than *cun*, but also 5 *fen* is one half which is
usually described by the character *ban* 半.

In the year 992, the official system for measures of weight consisted of
liang 兩, *qian* 錢, *fen*, *li*, *hao*, *si* and *hu* descending by tens [Needham
1959, p. 85]. In the thirteenth century, Yang Hui 楊輝 attached the names
fen, *li*, *hao*, *si* and *hu* freely to different measures such as *mu* 畝, *bu* 步 and
jin 斤 (see [Lam 1977, pp. 241–242]).

Names beyond the fifth decimal place can be found in the works of Qin
Jiushao 秦九韶 and Zhu Shijie 朱世傑. In his *Suanxue qimeng* 算學啓蒙
(Introduction to mathematical studies), Zhu Shijie compiled a table of names
for decimal fractions from 10^{-1} to 10^{-128} [Lam 1979, p. 8]. When Qin Jiushao
used rod numeral notations in his *Shushu jiu zhang* 數書九章 (Mathematical
treatise in nine sections) to express decimal fractions, he wrote the name of
the measure either above or below the digit occupying the units place; for
instance, 1.1446154 days was expressed as (see [Libbrecht 1973, p. 73]):

6.6   Densities

A short list pertaining to densities is given at the beginning of *Sun Zi suanjing*
after the tables of measures and the nomenclature of large numbers, but
there is no problem in the book on this subject. The list is as follows:

One *cun* 寸 cube of gold weighs 1 *jin* 斤,
one *cun* cube of silver weighs 14 *liang* 兩,
one *cun* cube of jade weighs 12 *liang*,
one *cun* cube of copper weighs $7\frac{1}{2}$ *liang*,
one *cun* cube of lead weighs $9\frac{1}{2}$ *liang*,
one *cun* cube of iron weighs 6 *liang*,
one *cun* cube of stone weighs 3 *liang*.

The Various Problems

7.1 The problems in *Sun Zi suanjing*

There are altogether 66 problems; 2 in Chapter 1, 28 in Chapter 2 and 36 in Chapter 3. Chapter 1 provides the basic tables and fundamental knowledge necessary for the understanding of the methods laid out in the solutions of the problems. Sun Zi started with the basics of arithmetic; he discussed the names of numbers and the formation of rod numerals which were the tools of computation. He then went on to describe how multiplication and division were performed, and illustrated the procedures in the only two problems of this chapter.

The subject on fractions which appeared in the first four problems of Chapter 2 has already been discussed in Sect. 4, and the extraction of a square root found in Probs. 19 & 20 of the same chapter has been discussed in Sect. 5.

It is clear that Sun Zi endeavoured to put as wide a selection of problems as possible in a primer. Some of them exhibit methods which we can classify, such as Rule of Three, Rule of False Position, the *fang cheng* 方程 method and the remainder method of indeterminate analysis. Some involve geometrical configurations for finding lengths, areas and volumes. Others such as those concerning partnerships involve the use of proportion, while some appear to be exercises to aid the understanding of the decimal place value concept. The remaining problems seem to be a potpourri which we have put under miscellany.

What is significant about these problems and many others earlier than these is that they are broadly similar to the outcrop of problems which arose in Europe after the Hindu-Arabic numeral system became an accepted mode of computation. (See, for instance, [Boncompagni 1857, Swetz 1987, pp. 37–175; de Hungaria 1499].) The latter occurrence was about a thousand years or more after the *Jiu zhang suanshu* was written. In both China and Europe, the aim of these problems was largely to provide illustrations and training in concepts and techniques which arose from the foundation of a numeral system that used place values with ten as base.

7.2   Rule of Three

Jiu zhang suanshu has probably the earliest general description of a widely used and much esteemed method in the commercial arithmetic of the ancient and medieval world. This method known as the Rule of Three involves the derivation of a fourth quantity from three known ones.

The rule is called *jin you* 今有 (lit. now there is) in *Jiu zhang suanshu* and the names of the three known quantities are: *suo qiu lü* 所求率 (the proportional value of the requested), *suo you shu* 所有數 (the amount of the given) and *suo you lü* 所有率 (the proportional value of the given). The fourth quantity, which is the amount of the requested, is derived from the use of the rule which is stated below [Qian ed. 1963, p. 114]:

> Multiply the proportional value of the requested (*suo qiu lü*) by the amount of the given (*suo you shu*) to form the *shi* 實 (dividend). Let the proportional value of the given (*suo you lü*) be the *fa* 法 (divisor). Divide the *shi* by the *fa* (*shi ru fa er yi* 實如法而一)。

This rule appeared to have originated from barter trading where one kind of goods was exchanged for another according to agreed or accepted proportional values. This is substantiated by the subject matter of the first thirty one problems of Chapter 2 in *Jiu zhang suanshu* and the meanings of the technical terms in the rule. The rule is employed to compute the exchange of foodstuff according to a list of proportional values placed at the beginning of the chapter. The given list of proportional values is shown below:

Millet (*su* 粟)	50
Coarse rice (*li mi* 糲米)	30
Polished rice (*bai mi* 粺米)	27
Refined rice (*zuo mi* 鑿米)	24
Imperial rice (*yu mi* 御米)	21
Ground wheat (*xiao zhe* 小麵)	13½
Ground barley (*da zhe* 大麵)	54
Cooked coarse rice (*li fan* 糲飯)	75
Cooked polished rice (*bai fan* 粺飯)	54
Cooked refined rice (*zuo fan* 鑿飯)	48
Cooked imperial rice (*yu fan* 御飯)	42
Big beans (*shu* 菽), small beans (*da* 荅), hemp (*ma* 麻) and wheat (*mai* 麥)	45
Paddy (*dao* 稻)	60
Fermented beans (*chi* 豉)	63
Cooked rice for evening meal (*sun* 飧)	90
Cooked big beans (*shou shu* 熟菽)	103½
Fermented grains (*nie* 糵)	175

A similar subject is found in four problems of *Sun Zi suanjing* (Ch. 2, Probs. 5–8); in fact, they are the same as the first four problems of Chapter 2 in *Jiu zhang suanshu*. While the method of each problem in *Jiu zhang suanshu* is described tersely in less than fifteen characters, that in *Sun Zi suanjing* is more expository.

Let us consider the first problem.

Now there is 1 *dou* of millet, find the equivalent amount of coarse rice.

Answer: 6 *sheng*.

The method in *Sun Zi suanjing* states:

Put down 1 *dou* or 10 *sheng* of millet. Multiply this by the proportional value of coarse rice, 30, to obtain 300 *sheng* which becomes the dividend (*shi*). The proportional value of millet, 50, becomes the divisor (*fa*). The answer is obtained on division.

Since we know how multiplication and division are performed with rods, it is easy to picture how the given method is transcribed into operations with rods.

In contrast with this method, that in *Jiu zhang suanshu* is very briefly stated in the following manner:

> To find coarse rice from millet: multiply by 3 and divide by 5.

The methods of the other thirty problems in the book are also stated in the same style. The same format was also used by Sun Zi when he listed the following with his collection of tables at the beginning of the book (p. 195).

> To find coarse rice from millet: multiply by 3 and divide by 5.
> To find millet from coarse rice: multiply by 5 and divide by 3.
> To find cooked [coarse] rice from coarse rice: multiply by 5 and divide by 2.
> To find cooked coarse rice from millet: multiply by 6 and divide by 4.
> To find coarse rice from cooked coarse rice: multiply by 2 and divide by 5.
> To find cooked [refined] rice from refined rice: multiply by 8 and divide by 4.

The next two problems of *Sun Zi suanjing*, Probs. 6 & 7, have fractions in the answers; in the working, the reader is reminded of the steps to be taken due to the presence of fractions. Prob. 7 (shown below) involves the proportional value of refined rice, which is 24, and the proportional value of millet, which is 50. As both numbers have the common factor 2, Sun Zi showed his preference for dividing each proportional value by 2 first before carrying out the Rule of Three.

> Now there are 4 *dou* 5 *sheng* of millet, find the equivalent amount of refined rice.
> Answer: 2 *dou* $1\frac{3}{5}$ *sheng*.
> Method: Put down 45 *sheng* of millet. Divide the proportional value of refined rice, 24, by 2 to give 12 and multiply [the amount of millet] by this to obtain 540 *sheng* which becomes the dividend (*shi*). Divide the proportional value of millet, 50, by 2 to give 25 which becomes the divisor (*fa*). Perform the division. The remainder [and the divisor] are divided by the *deng shu* 等數 (greatest common divisor) resulting in a fraction [of the simplest form].

There are two other problems in *Sun Zi suanjing* which are concerned with the exchange of commodities: Ch. 3, Probs. 11 & 16. The first problem involves an exchange of millet for beans and the second an exchange of millet for silk. In each case, the rate of exchange is stated in the problem. Exchanging goods for money is the subject of Ch. 3, Probs. 20 & 21; the former involves gold and *qian* 錢, and the latter brocade and *qian*. (*Qian* are ancient coins with a square hole in the centre.)

The brief methods of *Jiu zhang suanshu* offer a technique of computation which can calculate quickly the quantity of sought for foodstuff in exchange for a certain quantity of another foodstuff. Such a technique consists of a "multiply by this number and divide by that number" command, where the numbers have already been simplified or reduced wherever possible. Anyone who was involved in the trade of these foodstuff would naturally welcome such a prepared device, which would help the exchange to be performed swiftly and smoothly. In Chapter 1, Sun Zi gave this form of recitation for some of the more common foodstuff, but in the explanation of the methods in the problems of Chapter 2, he chose to show a detailed step by step application of the rule.

The general description of the rule in *Jiu zhang suanshu* and its application in a group of homogeneous problems at this early stage in the history of mathematics could well indicate how the rule was originally applied and then systematized. The present name of the rule was translated from the name given by the Hindu mathematician Brahmagupta (628), who described it as follows [Colebrooke 1817, p. 283]:

> In the rule of three, argument, fruit and requisition [are names of the terms]: the first and last terms must be similar. Requisition, multiplied by the fruit, and divided by the argument, is the produce.

It was also known by other names such as Golden Rule and Merchant's Rule [Smith 1925, pp. 484–488]. The importance and high regard that Europeans had for the rule for some three centuries since its introduction could be summed up in the following words of James Hodder [1672, p. 87]:

> The Rule of Three is commonly called, The Golden Rule; and indeed it might be so termed; for as Gold transcends all other Mettals, so doth this Rule all others in Arithmetick.

7.3 Geometrical configurations

Like most of the traditional mathematical texts, *Sun Zi suanjing* has quite a number of problems related to geometrical figures. They are collected here and discussed under the following subheadings.

7.3.1 Length

In Chapter 3, there are two problems involving length. They are self-explanatory and are quoted below; Prob. 18 is followed by Prob. 32.

> Now there is a piece of wood whose length is not known. If a rope is used to measure it, the rope has a remainder of 4 *chi* 5 *cun*. If the rope is bent [into two] and is then used to measure it, it is short of 1 *chi*. Find the length of the wood.
>
> Answer: 6 *chi* 5 *cun*.
>
> Method: Put down the remaining part of the rope, 4 *chi* 5 *cun*, and add the deficit, 1 *chi*, to give a total of 5 *chi* 5 *cun*. Double this to obtain 1 *zhang* 1 *chi* and subtract the remainder, 4 *chi* 5 *cun*, to get the answer.

> Now there is a canal 9 *li* long [with a line of] fish, each 3 *cun* long with its head touching the next fish. Find the number of fish.
>
> Answer: 54,000
>
> Method: Put down 9 *li*, multiply by 300 *bu* to obtain 2,700 *bu*. Next multiply by 6 *chi* to obtain 16,200 *chi*. Raise this to tens (*shang shi zhi* 上十之) to give 162,000 *cun*. Divide by the length of one fish, 3 *cun*, to get the answer.

7.3.2 Square and rectangle

Chapter 2 has three problems which deal with different aspects of the square. In Prob. 16, a rope, whose length is given, is arranged into a square, and finding a side of the square requires the division of the length by 4. Prob. 19 involves finding a side of a square from its given area; the detailed method of extracting the square root has already been discussed in Sect. 5.2.

Prob. 14 is stated below.

Now there is a square field with a mulberry tree at the centre. The distance of one corner [of the field] to the mulberry tree is 147 *bu*. Find the area of the field.

Answer: 1 *qing* 83 *mu* and an odd lot of 180 *bu*.

Method: Put down the distance from the corner to the mulberry tree, 147 *bu*, and double it to obtain 294 *bu*. Multiply by 5 to give 1470 *bu* and divide by 7 to obtain 210 *bu*. Multiply this by itself to obtain 44,100 *bu* and divide by 240 *bu* to get the answer.

The method uses an approximate ratio that the length of the diagonal of a square to its side is 7 to 5. The resulting answer is therefore only an approximation. At a time when an irrational number was non-existent, the ratio served a practical purpose of finding approximately one length from the other. In Chapter 1 (p. 193), the reader is given the following instructions:

[The ratio of] one side of a square to a diagonal is 5 to 7. Given a diagonal, to find a side: multiply by 5 and divide by 7. Given a side, to find a diagonal: multiply by 7 and divide by 5.

The above problem is also found in *Wu cao suanjing* 五曹算經 (The mathematical classic of five departments) [Qian ed. 1963, p. 414], which was written about the same time as *Sun Zi suanjing*. Yang Hui of the 13th century quoted the problem from *Wu cao suanjing* and suggested the method of squaring 147 *bu* and doubling the result to give the correct answer of 180 *mu* 18 *bu* (see [Lam 1977, p. 110]).

In Ch. 3, Prob. 5, a square chess board with 19 horizontal and 19 vertical lines is given and it is required to find the total number of chess pieces placed at the intersections of the lines. Prob. 22 of the same chapter gives a rectangle: 1,000 *bu* long and 500 *bu* wide. If there is a quail in every square *chi* and a small brown speckled bird in every square *cun*, we are asked to find the number of quails and speckled birds. In another problem (Ch. 2, Prob. 9) concerning the rectangular base of a house of length 6 *zhang* and width 3 *zhang*, we are asked to find the number of bricks required to lay the base, if every 5 pieces cover an area of 2 *chi*.

7.3.3 Circle and spherical segment

In *Sun Zi suanjing*, the ratio of the circumference of a circle to its diameter or, in other words, π, is always taken as 3. Ch. 2, Prob. 13 is concerned with a circular field where the circumference C and the diameter d are given. Three methods are put forward for finding the area A, namely, (i) $A = \frac{1}{2}C \times \frac{1}{2}d$, (ii) $A = C^2 \div 12$, (iii) $A = d^2 \times 3 \div 4$. In Ch. 2, Prob. 20, the area of a circle is given and, in order to find its circumference, we are asked to multiply the area by 12 and then find the square root of the product. Ch. 3, Prob. 33 mentions the distance between Chang' an and Luoyang and the circumference of a vehicle's wheel; we are asked to find the number of revolutions made by the wheel to cover the distance between the two cities.

The surface area of a spherical segment (*qiu tian* 丘田) in Ch. 2, Prob. 21 is given as a half of the circumference of the base multiplied by a half of the diameter of the sphere. This formula is first found in *Jiu zhang suanshu* (Ch. 1, Probs. 33 & 34) where a spherical segment is called *wan tian* 宛田. In the latter two problems, the character *zhou* 周, which means circumference, is prefixed by *xia* 下, which means base. (See fn. 13, p. 208). Liu Hui was aware of the inaccuracy of the method [Qian ed. 1963, p. 108]. Although Yang Hui pointed out that the method was incorrect, he did not give any alternative solution (see [Lam 1977, pp. 95–96]). *Wu cao suanjing* [Qian ed. 1963, p. 414] has a problem very similar to that in *Sun Zi suanjing*. A similar formula was used by ancient Indians (see [Gupta 1989, pp. 22–25]).

7.3.4 Volume

Millet is stored in different shaped cellars. The cellar is cylindrical in Ch. 2, Probs. 10 & 12, and the volume of the millet is obtained by squaring the given circumference, multiplying the depth and dividing by 12. The cellar of Ch. 2, Prob. 11 is in the shape of a rectangular parallelepiped. A drain of the same shape is given in Ch. 2, Prob. 18, where the width is 10 *zhang*, depth is 5 *zhang* and length is 20 *zhang*; it is required to find the number of cubes to fill up the drain when one cube has a volume of 1,000 *chi*.

Prob. 17, which precedes this question, requires the number of cubes to fill up an embankment in the shape of a prism whose cross section is a trapezium. If a = upper width, b = lower width, h = height and l = length, then the volume of the prism is given as $\frac{1}{2}(a+b)\,hl$. There are two other problems in the chapter which involve the same kind of volume. We are asked to find the manpower required to construct a city wall (Prob. 22) and a canal (Prob. 23), given that the work capacity of one person in autumn is 300 *chi*.

A simpler problem (Ch. 2, Prob. 15) gives a cube of wood of side 3 *chi*, and it is required to find the number of pillows to be carved from it, when each pillow requires a cube of side 5 *cun*. Ch. 3, Prob. 3 gives a conical pile of millet where its height and the circumference of its base are known. We are told to multiply the circumference by itself, multiply the product by the height and divide by 36.

7.4   Partnership and sharing

There are quite a number of problems on partnership and sharing in *Sun Zi suanjing*. The detailed description of some of the methods enables us to construct models on how they were solved through the use of rod numerals. At the same time, these models help us to perceive how mathematics evolved and grew through the medium of these numerals.

We shall discuss each problem in turn beginning with Ch. 2, Prob. 24.

Now there are 6,930 *qian* (coins) which are to be divided among 216 persons in 9 shares [as follows]: 81 persons get 2 shares each, 72 persons get 3 shares each, and 63 persons get 4 shares each. Find how much each person gets.

Answer: [For those with] 2 shares, each person gets 22 *qian*; [for those with] 3 shares, each person gets 33 *qian*; [and for those with] 4 shares, each person gets 44 *qian*.

Method: First put down 81 persons in the upper position, 72 persons in the next position and 63 persons in the lower position [i]. Multiply the upper numeral by 2 to obtain 162, multiply the next numeral by 3 to obtain 216, and multiply the lower numeral by 4 to obtain 252 [ii]. Add the numerals in the three positions to give 630 [iii], which becomes the divisor (*fa*). Next put down 6,930 *qian* in three positions [iv]. Multiply the upper numeral by 162 to obtain 1,122,660, multiply the middle numeral by 216 to obtain 1,496,880, and multiply the lower numeral by 252 to obtain 1,746,360 [v]. Each becomes the dividend (*shi*). Divide each by the divisor (*fa*),

630, to obtain 1,782 in the upper position, 2,376 in the middle position and 2,772 in the lower position [vi]. Divide each by the [corresponding] number of persons to get the answer [vii].

The stages of computation with rod numerals appear as follows:

⏚ Ι	Ι ⊥ ΙΙ	⊤ ☰
⊥ ΙΙ	ΙΙ — ⊤	
⊥ ΙΙΙ	ΙΙ ☰ ΙΙ	
[i]	[ii]	[iii]

⊥ ⫘ ☰	Ι — ΙΙ ☰ ⊤ ⊥
⊥ ⫘ ☰	Ι ☰ ⫘ ⊥ ⫘ ⏚
⊥ ⫘ ☰	Ι ⊥ ΙΙΙΙ ⊥ ΙΙΙ ⊥
[iv]	[v]

— ⊤ ⏚ ΙΙ	☰ ΙΙ
☰ ΙΙΙ ⊥ ⊤	☰ ΙΙΙ
☰ ⊤ ⊥ ΙΙ	☰ ΙΙΙΙ
[vi]	[vii]

In terms of Hindu-Arabic numerals, the above is as follows:

81	162	630	
72	216		
63	252		
[i]	[ii]	[iii]	

6930	1122660	1782	22
6930	1496880	2376	33
6930	1746360	2772	44
[iv]	[v]	[vi]	[vii]

The stages of computation are presented as they are depicted in the given method; no reasons are given for the steps taken. Steps [v] and [vii] may seem to us a little longwinded though the correct solution is obtained. In each stage of [i], [ii], [iv] to [vii], the upper, middle and lower positions act as notations standing for persons having two, three and four shares respectively.

The next problem (Ch. 2, Prob. 25) concerns the distribution of tangerines to five noblemen according to their ranks.

> Now there are five noblemen of five different ranks sharing a total of 60 tangerines. If each person has 3 tangerines more [than the next person of a lower rank], find how many tangerines each man receives.
>
> Answer: *Gong* 公 (duke) gets 18, *Hou* 侯 (marquis) gets 15, *Bo* 伯 (earl) gets 12, *Zi* 子 (viscount) gets 9, and *Nan* 男 (baron) gets 6.
>
> Method: First put down in the bottom position 3 tangerines, which is the amount to be added for each successive rank. [Put down] 6 tangerines in the next position [above], 9 tangerines in the next position, 12 tangerines in the next position, and 15 tangerines in the top position [i]. Add them to obtain 45 [ii]. Subtract this from 60 and divide the remainder by the number of persons to give 3 tangerines for each person [iii]. Add this to each [of the numerals] before they were added up (*bu bing zhe* 不并者) to obtain 18 in the top position which is *Gong's* share, 15 in the next position [below] which is *Hou's* share, 12 in the next position which is *Bo's* share, 9 in the next position which is *Zi' s* share, and 6 in the bottom position which is *Nan's* share [iv].

The main stages of computation on the board with rods are constructed below and these are followed by an interpretation in Hindu-Arabic numerals.

| [i] | [ii] | [iii] | [iv] |

15	45	3	18
12			15
9			12
6			9
3			6
[i]	[ii]	[iii]	[iv]

As in the previous problem, the positions of the rows in [i] and [iv] have an implicit meaning; in this case, they correspond to the ranks of the five noblemen.

The next problem (Prob. 26) is also concerned with laying down numerals in positions which stand for different persons.

> Now there are three persons A, B and C who hold certain sums of money. A says to B and C, "If one half of each of your money is added to mine, the result is 90." B says to A and C, "If one half of each of your money is added to mine, the result is 70." C says to A and B, "If one half of each of your money is added to mine, the result is 56." How much money does each of the three men hold originally?
> Answer: A 72, B 32 and C 4.
> Method: First put down in [three] positions the amounts declared by the three persons [i], and multiply each by 3 obtaining 270 for A, 210 for B and 168 for C [ii]. Halving each yields 135 for A, 105 for B and 84 for C [iii]. Once again put down 90 for A, 70 for B and 56 for C [iv], and halve each of them [v]. Subtract [the latter] A and B from [the previous] C, [the latter] A and C from [the previous] B, and [the latter] B and C from [the previous] A, to yield the original amounts of money held by each person [vi].

The stages of computation using rod numerals as described in the method are shown below:

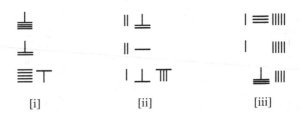

[i]	[ii]	[iii]

| [iv] | [v] | [vi] |

The above translated into Hindu-Arabic numerals are as follows:

90	270	135	90	45	72
70	210	105	70	35	32
56	168	84	56	28	4
[i]	[ii]	[iii]	[iv]	[v]	[vi]

In each of the above columns, the numeral in the first row is related to A, the numeral in the second row to B, and the numeral in the third row to C. First the given numerals 90, 70 and 56 are laid down in their respective positions (see [i]), next the numerals are multiplied by 3 (see [ii]), and their products are divided by 2 (see [iii]). The given numerals are once again laid down (see [iv]) and then they are halved (see [v]). Finally we are asked to perform the following steps:

1. Subtract the numerals of A and B in [v] from that of C in [iii], i.e., 45 and 35 from 84, to give 4 for C (see [vi]).
2. Subtract the numerals of A and C in [v] from that of B in [iii], i.e., 45 and 28 from 105, to give 32 for B (see [vi]).
3. Subtract the numerals of B and C in [v] from that of A in [iii], i.e., 35 and 28 from 135, to give 72 for A (see [vi]).

No explanation is given for all the steps. Whatever reasons that are now put forward would be speculative. Since this is an algebraic problem, let us solve it according to the concept and notation that we are familiar with. The problem involves three linear equations in three unknowns.

Let a, b and c be the amounts of money that A, B and C originally hold. Then

$$a + \tfrac{1}{2}b + \tfrac{1}{2}c = 90 \qquad (1)$$

$$\tfrac{1}{2}a + b + \tfrac{1}{2}c = 70 \qquad (2)$$

$$\tfrac{1}{2}a + \tfrac{1}{2}b + c = 56. \qquad (3)$$

Multiply equation (3) by 3 and divide by 2,

$$\tfrac{3}{4}a + \tfrac{3}{4}b + \tfrac{3}{2}c = 84. \qquad (4)$$

Divide equation (1) by 2,

$$\tfrac{1}{2}a + \tfrac{1}{4}b + \tfrac{1}{4}c = 45. \qquad (5)$$

Divide equation (2) by 2,

$$\tfrac{1}{4}a + \tfrac{1}{2}b + \tfrac{1}{4}c = 35. \qquad (6)$$

Subtract equations (5) and (6) from equation (4) to give $c = 4$. In a similar manner, we obtain $a = 72$ and $b = 32$.

In Ch. 3, Prob. 1, the varying amounts of grains which 9 households have to pay for land tax and the transportation of the grains are stated. Given the cost of the transport, we are asked to find the amount of grains that each household pays for the land tax alone.

Now there are 9 households A, B, C, D, E, F, G, H and I who collectively send their land tax [in kind]. A pays 35 *hu* [of grains], B 46 *hu*, C 57 *hu*, D 68 *hu*, E 79 *hu*, F 80 *hu*, G 100 *hu*, H 210 *hu* and I 325 *hu*. All 9 households send a total of 1,000 *hu* to pay their land tax. Out of this 200 *hu* is deducted for transport, find [the actual tax] each household pays.

Answer: A 28 *hu*, B 36 *hu* 8 *dou*, C 45 *hu* 6 *dou*, D 54 *hu* 4 *dou*, E 63 *hu* 2 *dou*, F 64 *hu*, G 80 *hu*, H 168 *hu* and I 260 *hu*.

Method: Put down A's payment of 35 *hu*, multiply by 4 to obtain 140 *hu* and divide by 5 to give 28 *hu*. B pays 46 *hu*, multiply by 4 to obtain 184 *hu* and divide by 5 to give 36 *hu* 8 *dou*. C pays 57 *hu*, multiply by 4 to obtain 228 *hu* and divide by 5 to give 45 *hu* 6 *dou*. D pays 68 *hu*, multiply by 4 to obtain 272 *hu* and divide by 5 to give 54 *hu* 4 *dou*. E pays 79 *hu*, multiply by 4 to obtain 316 *hu* and divide by 5 to give 63 *hu* 2 *dou*. F pays 80 *hu*, multiply by 4 to obtain 320 *hu* and divide by 5 to give 64 *hu*. G pays 100 *hu*, multiply by 4 to obtain 400 *hu* and divide by 5 to give 80 *hu*. H pays 210 *hu*, multiply by 4 to obtain 840 *hu* and divide by 5 to give 168 *hu*. I pays 325 *hu*, multiply by 4 to obtain 1,300 *hu* and divide by 5 to give 260 *hu*.

The computation for each household is shown below in [i], [ii] and [iii] followed by an interpretation in Hindu-Arabic numerals.

[i]	[ii]	[iii]
35	140	28
46	184	36.8
57	228	45.6
68	272	54.4
79	316	63.2
80	320	64
100	400	80
210	840	168
325	1300	260
[i]	[ii]	[iii]

In [i], the contributions of the 9 households, A to I, are laid down in 9 rows and the quantities are in *hu* units. Next the numerals are multiplied by 4 (see [ii]) and then divided by 5 to give the solution (see [iii]). In [ii], the quantities are in *hu* as stated in the text, and after a division by 5, the quantities may be in *hu* and *dou* units. The first case for A is straightforward: 140 *hu* divided by 5 gives 28 *hu*. In the case for B, we are told that 184 *hu* divided by 5 gives 36 *hu* 8 *dou*. How did the ancient computer arrive at the result? There are three possible ways, and since no further explanation is given we are not sure which is the correct one. The possible methods are:

1. Since 1 *hu* = 10 *dou*, the computer is aware of the decimal fraction concept, and a division of 184 by 5 through the medium of rod numerals results in the placement of rods shown in the second row of [iii].
2. The computer divides 184 by 5 to obtain $36\frac{4}{5}$ *hu* (i.e., 𝍫𝍥
𝍣
𝍤), and

   $\frac{4}{5}$ *hu* is 8 *dou*.
3. The computer divides 184 by 5 to obtain 36 *hu* and a remainder of 4 *hu*; the latter is converted to 40 *dou* and then divided by 5 to obtain 8 *dou*. The final numerals on the board may then be 𝍫𝍥 𝍧.

It is not necessary to discuss Ch. 3, Probs. 8, 10 & 23 as they are straightforward. In Ch. 3, Prob. 9, instead of asking the reader to multiply 36, 454 families by 40 *liang* (= 2 *jin* 8 *liang*), Sun Zi asked him to "raise this to tens (*shang shi zhi* 上十之) to obtain 364,540" and then to "multiply by 4 to give 1,458,160 *liang*". This indicates that Sun Zi was aware that the multiplication of a numeral by 10 meant simply raising its digits by one rank. This entails the shifting of the rod digits on the computing board to the next place on the left, so that 𝍢𝍮𝍣𝍭𝍣 would appear as 𝍫𝍥𝍬𝍤𝍬 . The phrase "raise this to tens (*shang shi zhi*)" is also found in the following problems: Ch. 3, Probs. 22, 25 and 32. In Prob. 22, it is incorrectly used.

The change in the ranks of the digits of a numeral when multiplied by numbers in powers of ten has been recorded by Xiahou Yang 夏侯陽. We give below a literal translation of the passage [Qian ed. 1963, p. 559].

Multiply by 10: add 1 rank. Multiply by 100: add 2 ranks. Multiply
by 1,000: add 3 ranks. Multiply by 10,000: add 4 ranks. Divide by
10: move back 1 rank. Divide by 100: move back 2 ranks. Divide
by 1,000: move back 3 ranks. Divide by 10,000: move back 4 ranks.

Ch. 3, Prob. 15 (see below) appears to us to involve two equations in
two unknowns, and we can only hypothesize on the reasoning behind the
method presented.

Now when every 3 persons share a cart, there are 2 carts empty. If
every 2 persons share a cart, 9 persons have to walk. Find the number
of persons and the number of carts.
Answer: 15 carts; 39 persons.
Method: Put down 2 carts, multiply by 3 to give 6, add 9, which is
the number of persons who have to walk, to obtain 15 carts. To find
the number of persons, multiply the number of carts by 2 and add 9,
which is the number of persons who have to walk.

Let x be the number of persons and y the number of carts, then our two
equations are

$y = (x \div 3) + 2$, which is equivalent to $3y = x + 6$, and $2y = x - 9$.
Solving them gives $y = 15$.

The next problem (Ch. 3, Prob. 17) involves the solution of a simple
linear equation.

Now there was a woman washing bowls by the river. An officer
asked, "Why are there so many bowls?" The woman replied, "There
were guests in the house." The officer asked, "How many guests
were there?" The woman said, "I don't know how many guests
there were; every 2 persons had [a bowl of] rice, every 3 persons
[a bowl of] soup and every 4 persons [a bowl of] meat, 65 bowls
were used altogether."
Answer: 60 persons.
Method: Put down 65 bowls, multiply by 12 to obtain 780 and
divide by 13 to get the answer.

Let x be the number of persons, then $(x \div 2) + (x \div 3) + (x \div 4) = 65$, resulting in $x = (65 \times 12) \div 13$.

The method of Ch. 3, Prob. 19 is also obvious as shown below.

> Now there is an unknown quantity of rice in a container. If the first person takes a half of it, the second person takes a third [of the remainder] and the last person takes a quarter [of what is left], the remaining amount of rice is 1 *dou* 5 *sheng*. Find the original amount of rice.
>
> Answer: 6 *dou*.
>
> Method: Put down the remaining amount of rice, 1 *dou* 5 *sheng*, multiply by 6 to obtain 9 *dou* and divide by 2 to give 4 *dou* 5 *sheng*. Multiply by 4 to obtain 1 *hu* 8 *dou* and divide by 3 to get the answer.

Lastly Ch. 3, Prob. 30 is concerned with the proportion that the owners of chicks, hens and roosters have to pay the millet seller according to the amount of grains consumed by each fowl. The problem and method are stated below.

> Now there are 3 fowls which peck a total of 1001 grains of millet. If the chick pecks 1 grain, the hen pecks 2 and the rooster pecks 4, find the [proportional] amounts the owners of the three kinds of fowls have to pay the millet seller.
>
> Answer: The owner of the chick pays 143, the owner of the hen pays 286, the owner of the rooster pays 572.
>
> Method: Put down 1001 grains as the dividend (*shi*). Next take the sum of 7 grains pecked by the 3 fowls as the divisor (*fa*). Divide to obtain 143 grains, which is the amount the chick owner has to pay. Doubling this gives the amount the hen owner has to pay and [doubling again] gives the amount the rooster owner has to pay.

7.5   The remainder problem

The famous and oldest example of the remainder problem is in Ch. 3, Prob. 26.

Now there are an unknown number of things. If we count by threes, there is a remainder 2; if we count by fives, there is a remainder 3; if we count by sevens, there is a remainder 2. Find the number of things.
Answer: 23.
Method: If we count by threes and there is a remainder 2, put down 140. If we count by fives and there is a remainder 3, put down 63. If we count by sevens and there is a remainder 2, put down 30. Add them to obtain 233 and subtract 210 to get the answer. If we count by threes and there is a remainder 1, put down 70. If we count by fives and there is a remainder 1, put down 21. If we count by sevens and there is a remainder 1, put down 15. When [a number] exceeds 106, the result is obtained by subtracting 105.

The second part of the method from "If we count by threes and there is a remainder 1, put down 70", and so forth, appears to be an explanation of how the figures 140, 63 and 30 in the first part were derived. One would imagine that the figures 70, 21 and 15 were first displayed on the board (see [i]). Each numeral was then multiplied by its corresponding given "remainder": 2, 3 and 2 respectively (see [ii]); and the products were added up (see [iii]). Finally 210 was subtracted from the sum to obtain the answer (see [iv]).

[i]   [ii]   [iii]   [iv]

No explanations were offered on how 70, 21 and 15 in [i] were derived nor why a multiple of 105 must be subtracted from the sum in [iii]. This problem has since evolved into what is now known as the Chinese Remainder theorem. With the advantage of being able to use a sophisticated set of notations, we shall now present the explanation which Sun Zi omitted. This explanation is found in most elementary text book on number theory (see, for instance, [Dickson 1957, pp. 11–12]). The theorem is stated as follows:

If $m_1, \ldots, m_t$ are relatively prime in pairs, there exist integers x for which simultaneously

$$x \equiv a_1 \ (\text{mod } m_1), \ldots, x \equiv a_t \ (\text{mod } m_t) \qquad (1).$$

All such integers x are congruent modulo $m = m_1 m_2 \ldots m_t$.

The explanation is as follows:

Set $m = m_1 M_1 = \ldots = m_t M_t$.
Then M_1 is prime to $m_1, \ldots, M_t$ is prime to m_t.
Hence, we can determine integers $b_1, \ldots, b_t$ such that
$$M_1 b_1 \equiv 1 \ (\text{mod } m_1), \ldots, M_t b_t \equiv 1 \ (\text{mod } m_t).$$
Then congruences (1) are all satisfied if
$$x = M_1 b_1 a_1 + \ldots + M_t b_t a_t.$$
In fact, since $M_2, \ldots, M_t$ are all divisible by m_1,
$$x \equiv M_1 b_1 a_1 \equiv a_1 \ (\text{mod } m_1).$$
Similarly, $x \equiv M_t b_t a_t \equiv a_t \ (\text{mod } m_t)$.
It is obvious that all $x = M_1 b_1 a_1 + \ldots + M_t b_t a_t \ (\text{mod } m)$ also satisfy (1).

In the above problem of Sun Zi, $m_1 = 3$, $m_2 = 5$, $m_3 = 7$; $a_1 = 2$, $a_2 = 3$, $a_3 = 2$. $M_1 b_1$, $M_2 b_2$ and $M_3 b_3$ are 70, 21 and 15 respectively (see [i]) and $x = M_1 b_1 a_1 + M_2 b_2 a_2 + M_3 b_3 a_3 = 140 + 63 + 30 = 233$ (see [ii] & [iii]). The answer is given in its lowest term, where $x \equiv 23 \ (\text{mod } 105)$.

It is fortunate that we have knowledge of this problem from Sun Zi's book. The solution was employed by the ancient astronomers to perform complicated computations of the calendar, but their methods have not survived to the present. (See [Li & Du 1987, p. 94].) What we can say is that the remainder problem had been extended and developed, and by the 13th century Qin Jiushao's work showed a sophisticated method of solution which also tackled the difficult case where the moduli were not relatively prime. (See [Libbrecht 1973, pp. 328–366, Li & Yuan 1983].)

7.6   Rule of False Position

Let us look at Ch. 3, Prob. 29 which states:

Now there are 100 deer [being distributed] in a city. If one household has one deer there is a remainder, and if the remainder is again [being distributed] such that every three households share a deer, then nothing is left. Find the number of households in the city. Answer: 75 households.

To solve this problem, our present knowledge of algebra enables us to formulate a simple equation in terms of the unknown x, which in this case is the number of households. The equation appears as $3(100 - x) = x$, and the solution is expeditiously derived. How would this problem be solved by one whose knowledge was limited to the fundamental operations of arithmetic using either rod numerals or the Hindu-Arabic numerals? Sun Zi wrote the following solution which was to be derived through the manipulation of rod numerals:

> Method: Use [the method of] *ying bu zu* 盈不足 (surplus and deficit). If there are 72 households, there is a surplus of 4 deer. If there are 90 households, there is a deficit of 20 deer. Put down 72 in the upper position of the right column and the surplus, 4, in the lower position of the right column. Put down 90 in the upper position of the left column and the deficit, 20, in the lower position of the left column. Cross-multiply (*wei cheng* 維乘) and add the results to form the dividend (*shi*). Add the surplus and deficit to form the divisor (*fa*). Divide to get the answer.

The Chinese called the method *ying bu zu* meaning "surplus and deficit" which is equivalent to what was known in the West as the Rule of False Position. (See [Smith 1925, pp. 437–442].) The method reflected an early stage in the evolution of man's intellect before he was able to assign written symbols to stand for objects and to operate with them. Sometimes, as in the above problem, the method was used to solve what we now call a simple linear equation of the form $ax + b = 0$. The justification of the ancient method is apparent when we assume knowledge of this equation. Two guesses g_1 and g_2 for x are presented giving values f_1 and f_2 for $ax + b$. If a value is positive, it represents a "surplus" and if it is negative, it represents a "deficit". The solution for x is given as $(f_1 g_2 - f_2 g_1)/(f_1 - f_2)$.

Sun Zi's method explains clearly where the rod numerals are placed on the counting board: 72 is put in the upper position of the right column with the corresponding surplus, 4, below it, and 90 is put in the upper position of the left column with the corresponding deficit, 20, below it. Thus the placement of the rod numerals is as follows:

$$\underset{\equiv}{\underline{\perp}} \qquad \underset{\text{IIII}}{\overset{\perp}{\underline{\quad}}} \text{ II}$$

In Hindu-Arabic numerals, the above appears as

90	72
20	4

The placing of the numerals in this form is a significant step in the initial development of arithmetic and algebra. Inevitably, the concept of cross-multiplication arises; the Chinese term for this is *wei cheng* 維乘. Secondly, the positions of the numerals symbolize what the numerals stand for; in the above, the upper positions stand for the proposed values and the lower positions stand for their corresponding surplus or deficit. After positioning the rod numerals on the counting board, we are asked to "cross-multiply and add the results". This gives $72 \times 20 + 90 \times 4 = 1,800$, which is called *shi*. The sum of the surplus and deficit, i.e., $4 + 20 = 4$, is called *fa*. The answer is obtained by dividing the *shi* by the *fa*.

How did the *ying bu zu* method originate? The answer is unexpectedly simple: from problems dealing with surplus and deficit. The *Jiu zhang suanshu* has an entire chapter (Ch. 7) of 20 problems devoted to this subject, and the first eight problems are different from the other twelve as they are specifically concerned with surpluses and/or deficits. (See [Lam 1974].) Let us look at the first problem [Qian ed. 1963, p. 205]:

> Now there is a certain number of persons buying goods. If each person pays 8 there is a surplus (*ying*) of 3, and if each person pays 7 there is a deficit (*bu zu*) of 4. Find the number of persons and the cost of the goods.
>
> Answer: 7 persons, the goods cost 53.

This is followed by three other similar problems which involve a surplus and a deficit. No method is given individually for any of the questions, but a general method is stated after Prob. 4. The first part of the method states:

Ying bu zu. The method says: Put down the proposed values (*suo chu lü* 所出率) and below each, the surplus (*ying*) or deficit (*bu zu*). Cross-multiply (*wei cheng*) [the latter] and the proposed values (*suo chu lü*), add [the products] and call [the sum] *shi*. Add the surplus (*ying*) and deficit (*bu zu*) and call [the sum] *fa*. Divide the *shi* by the *fa* (*shi ru fa er yi*). If there are fractions, "communicate" (*tong* 通) them.

The procedure is the same as that of Sun Zi's method discussed above. If we were to apply this to any of the four problems, the result obtained is the exact amount that each person pays for the goods. However, this is not asked in the problems, which require the number of persons and the cost of the goods. The above passage now continues to give the procedure of finding these unknowns:

In [cases] where the surplus (*ying*) and deficit (*bu zu*) are mutually connected with persons buying goods, put down the proposed values (*suo chu lü*) and subtract the smaller from the larger. Divide (*yue* 約) the *shi* and the *fa* by the remainder. [Division of] *shi* gives the cost of the goods and [division of] *fa* gives the number of persons.

Note that the *shi* and *fa* mentioned here refer to the terms defined earlier in the passage. That is, the *shi* refers to the sum of the products after cross-multiplication has taken place, and the *fa* refers to the sum of the surplus and deficit. Each of the terms is now divided by the difference between the proposed values to give the cost of the goods and the number of persons respectively.

Sun Zi suanjing has a similar problem (Ch. 2, Prob. 28):

Now there is a gang of robbers who stole an unknown quantity of thin silk from a warehouse. In the distribution of the silk among themselves, it is heard that if each person is given 6 *pi*, there is a surplus of 6 *pi*, and if each person is given 7 *pi*, there is a deficit of 7 *pi*. Find the number of persons and the amount of thin silk.
Answer: 13 robbers and 84 *pi* of thin silk.
Method: First put down on the upper right, the amount each person gets, 6 *pi*, and on the lower right, the surplus, 6 *pi*. After this put down on the upper left, the amount each person gets, 7 *pi*, and on

the lower left, the deficit, 7 *pi*. Cross-multiply (*wei cheng*) and add the results to obtain the quantity of thin silk. Add the surplus and deficit to obtain the number of persons.

Though the answers are correct, the method illustrates an obvious weakness in its style as an application of the same technique to a similar problem may not yield the correct answers. This is because each result has yet to be divided by the difference of the proposed values, which in this case happens to be 1; the explanation of this is omitted in the method.

Besides the above two problems, there are no others on the *ying bu zu* method in *Sun Zi suanjing*. The *Jiu zhang suanshu* has a discussion of two other possibilities of the method: (i) two surpluses (*liang ying* 兩盈) or two deficits (*liang bu zu* 兩不足), and (ii) surplus and exactness (*ying, shi zu* 適足) or deficit and exactness (*bu zu, shi zu*). (See [Lam & Shen 1989, pp. 110–112].)

The description of the first case follows the same pattern as that for *ying bu zu* (surplus and deficit), and we translate it below so that the reader can compare them [Qian ed. 1963, p. 208].

> *Liang ying, liang bu zu*. The method says: Put down the proposed values (*suo chu lü*) and below each, the surpluses (*ying*) or deficits (*bu zu*). Cross-multiply (*wei cheng*) [the latter] and the proposed values (*suo chu lü*), subtract the smaller [product] from the larger and call the remainder *shi*. In the two surpluses (*liang ying*) or the two deficits (*liang bu zu*), subtract the smaller from the larger and call the remainder *fa*. Divide the *shi* by the *fa* (*shi ru fa er yi*). If there are fractions, "communicate" (*tong*) them. In [cases] where the two surpluses (*liang ying*) or the two deficits (*liang bu zu*) are mutually connected with persons buying goods, put down the proposed values (*suo chu lü*) and subtract the smaller from the larger. Divide (*yue*) the *shi* and the *fa* by the remainder. [Division of] *shi* gives the cost of the goods and [division of] *fa* gives the number of persons.

Since there is a surplus and a deficit in the *ying bu zu* method, the *shi* is obtained by adding the products after cross-multiplication. In the other method, since there are either two surpluses or two deficits, the *shi* is obtained by subtracting the products. Based on the same reasons, the *fa* in the *ying bu zu* method is obtained by adding the surplus and the deficit, and in the other method, it is obtained by subtracting the smaller quantity from the larger of the two surpluses or the two deficits.

Early this century, Qian Baocong [1927; 1964, pp. 34–41] and Zhang Yinlin 張蔭麟 [1927] drew attention to the fact that the method known in Europe as the Rule of False Position was the same as the *ying bu zu* method. Known to the Arabs as *hisāb al-khatā' ain*, it can be found in the works of al-Khwārizmī (c.825), Qustā ibn Lūqa al-Ba' albakī (d.922) and other writers [Smith 1925, p. 437; Needham 1959, p. 118]. The rule adopted various names in Europe such as *elchataym* (Fibonacci, 13th century), *el cataym* (Pacioli, 15th century), *regola helcataym* (Tartaglia, 16th century) and *regole del cattaino* (Pagnani, 16th century).

As the rule could be applied mechanically to a variety of problems, it proved to be popular and useful at a time when man was still unable to formulate simple equations in terms of written symbols. It was faster and easier to use the rule to obtain the answer than to embark on the difficult task of devising a new method to solve the problem. Smeur [1978, p. 68] shared similar views on the use of the rule when he said, "Its importance is that it is not necessary to frame an equation first but that one can calculate at once with any number after which the correct result can be found from the deviations. This really is important for it is well-known that for the less experienced, when meeting with quite simple problems as mentioned, the difficulty is not to solve an equation (which can easily be learned) but the main difficulty is first to frame such an equation by reading the problem very carefully."

7.7　*Fang cheng* method

This method translated as "rectangular tabulation", (lit. square procedure) solves a set of simultanous linear equations. Chapter 8 of *Jiu zhang suanshu* is concerned with this method; it has eighteen problems, eight of which (Probs. 2, 4–6, 7, 9–11) involve two equations in two unknowns, six (Probs. 1, 3, 8, 12, 15, 16) with three equations in three unknowns and two (Probs. 14, 17) with four equations in four unknowns. Prob. 18 involves five equations in five unknowns and Prob. 13 is indeterminate with five equations in six unknowns.

Sun Zi suanjing has only one problem that uses this method. The problem (Ch. 3, Prob. 28) somewhat similar to Prob. 10 of *Jiu zhang suanshu* is stated as follows:

> Now there are 2 persons, A and B, each of whom holds an unknown amount of money. If A gets one half of B's [money], he has a total of 48. If B gets two thirds (*da ban* 大半) of A's [money], he also has a total of 48. Find how much money each person holds.
> Answer: A holds 36, B holds 24.
> Method: Use [the method] of *fang cheng* 方程 (rectangular tabulation) to solve. Put down in the right column 2 for A, 1 for B and 96 for money. Put down in the left column 2 for A, 3 for B and 144 for money [i]. Multiply the left column by the 2 of the right column to obtain 4 for A in the upper position, 6 for B in the middle position, and 288 for money in the lower position [ii]. Use the right column to subtract the left column twice so that in the left column, the top position is empty, the middle position has remainder 4 for B, which becomes the divisor (*fa*), and the lower position has remainder 96 for money, which becomes the dividend (*shi*) [iii]. With the divisor (*fa*) above and the dividend (*shi*) below, [divide] to obtain 24, which is the money B holds. Subtract this from the 96 in the lower position of the right column to give a remainder of 72, which becomes the dividend (*shi*). Let the 2 for A in the upper position of the right column be the divisor (*fa*). With the divisor (*fa*) above and the dividend (*shi*) below, [divide] to obtain 36, which is the money A holds.

The stages of computation with rod numerals would very likely be as follows:

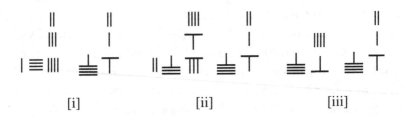

[i]   [ii]   [iii]

In Hindu-Arabic numerals, the above appears as:

2	2	4	2	0	2
3	1	6	1	4	1
144	96	288	96	96	96

| [i] | | [ii] | | [iii] | |

The problem involves two equations of the form

$$2x + y = 96$$
$$2x + 3y = 144.$$

The Chinese solved the problem by writing the numerals in two columns, (see [i]), so that each column represented an equation. The positions in the first and second rows stood for the respective portions of A's and B's money, and those in the third row represented the total amount of money. In order to eliminate the numeral in the top position of the left column, the numerals in the left column were multiplied by the number in the top position of the right column (see [ii]), and then the left column successively subtracted the right one until there were no numeral in its first row (see [iii]). This meant that the column now represented only B's portions and the corresponding amount of money. After one portion of B's money was derived, the final step was to find one portion of A's money from the right column.

The method showed that the Chinese had evolved a general procedure to solve a pair of linear equations in two unknowns, and it could be mechanically applied to obtain the answers. Sun Zi gave us an insight into the general method when he showed every step. The two numbers occupying the first row (see [i]) happened to be the same, and the working could have been shortened by omitting the step of multiplying the left column by the number in the first row of the right column. The left column could straightaway subtract the right one, thus eliminating the numeral in its first row; this, however, would not be a general method.

The *fang cheng* method in *Jiu zhang suanshu* is shown in only one problem despite the fact that there are eighteen problems involved with simultaneous linear equations. This problem (Ch. 8, Prob. 1) is concerned with the solution of three equations in three unknowns. A study of the method reveals that it is an extension of the method used by Sun Zi to solve two equations in two unknowns. The problem and method [Qian ed. 1963, pp. 221–222] are translated here followed by our construction of the stages of computation with rod numerals.

Now there are 3 bundles of top grade cereal, 2 bundles of medium grade cereal and 1 bundle of low grade cereal, which yield 39 *dou* [of grains] as *shi*; 2 bundles of top grade cereal, 3 bundles of medium grade cereal and 1 bundle of low grade cereal yield 34 *dou* as *shi*; 1 bundle of top grade cereal, 2 bundles of medium grade cereal and 3 bundles of low grade cereal yield 26 *dou* as *shi*. Find the measure [of grains] in each bundle of the top, medium and low grade cereal. Answer: One bundle of top grade cereal [has] $9\frac{1}{4}$ *dou*, one bundle of medium grade cereal [has] $4\frac{1}{4}$ *dou*, and one bundle of low grade cereal [has] $2\frac{3}{4}$ *dou*.

Method of *fang cheng*: Put down 3 bundles of top grade cereal, 2 bundles of medium grade cereal 1 bundle of low grade cereal and 39 *dou* as *shi* in a column on the right. Set up the columns in the centre and on the left in the same way as the column on the right [i]. Take the [number representing] top grade cereal in the right column to multiply all [numbers] in the central column [ii], and then use [the method] of direct subtractions (*zhi chu* 直除) [iii]. Once again multiply [the numbers] in the next column [that is, the left column, by the number representing top grade cereal in the right column] [iv], and then use [the method of] direct subtractions (*zhi chu*) [v]. Next multiply all [the numbers in] the left column by the remaining [number representing] medium grade cereal in the central column [vi], and then use [the method of] direct subtractions [vii]. The left column has the remaining number [representing] low grade cereal. The *fa* (divisor) is above and the *shi* (dividend) below; the *shi* here is the *shi* for low grade cereal. To find [the measure for] medium grade cereal, multiply the *shi* in the central column by the *fa* [of the left column], and subtract the *shi* for low grade cereal [viii]. The remainder is divided by the number of bundles of medium grade cereal [in the central column], yielding the *shi* for medium grade cereal [ix]. To find [the measure for] top grade cereal, once again multiply the *shi* in the right column by the *fa* [of the left column], and subtract [the respective] *shi* for low and medium grades [x]. The remainder is divided by the number of bundles of top grade cereal [in the right column], yielding the *shi* for top grade cereal [xi]. The *shi* for all [grades] are each divided by the *fa* to yield the measures [per bundle of the respective grades] [xii].

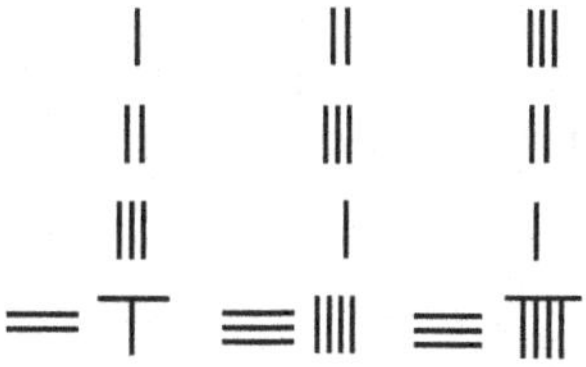

[i]

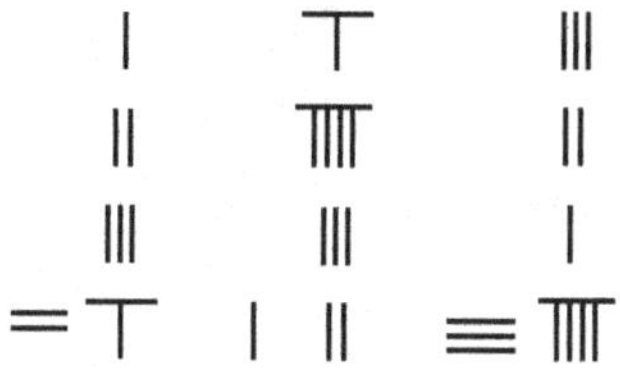

[ii]

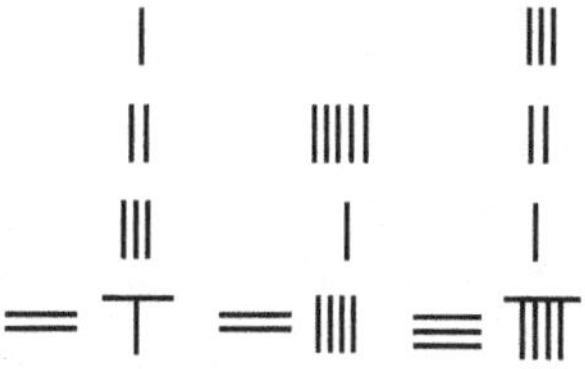

[iii]

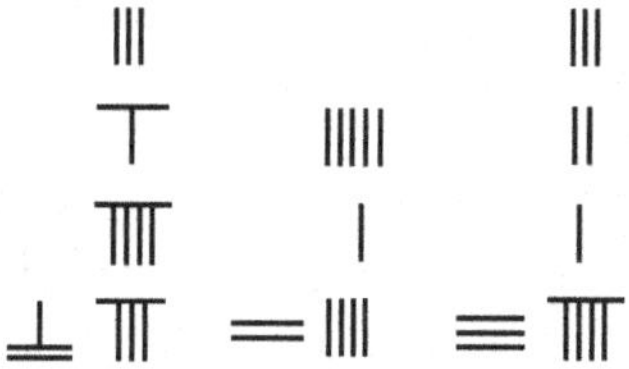

[iv]

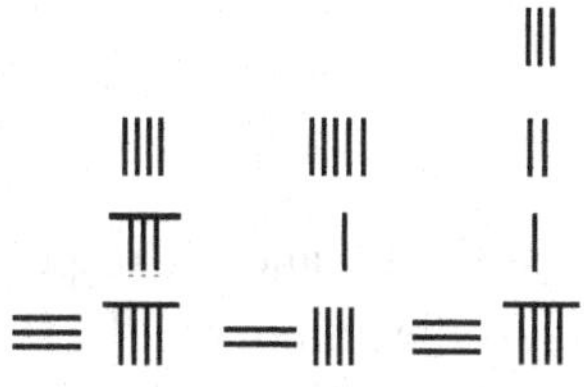

[v]

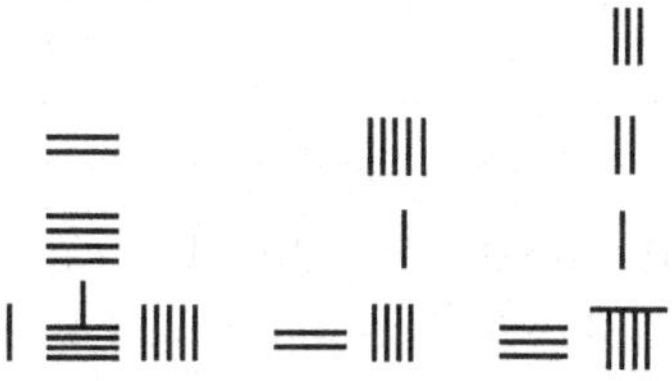

[vi]

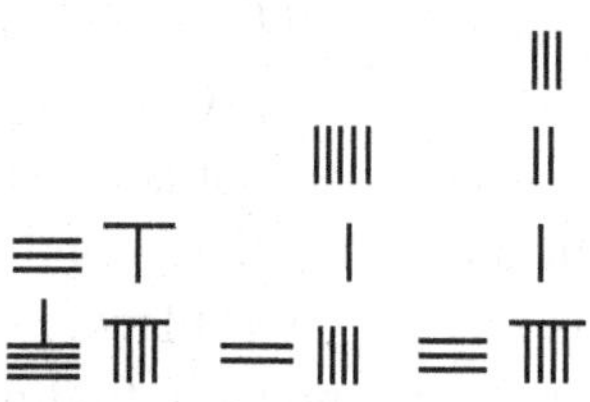

[vii]

Translated into Hindu-Arabic numerals, the above appears as follows:

1	2	3
2	3	2
3	1	1
26	34	39

[i]

1	6	3
2	9	2
3	3	1
26	102	39

[ii]

1	0	3
2	5	2
3	1	1
26	24	39

[iii]

3	0	3
6	5	2
9	1	1
78	24	39

[iv]

0	0	3
4	5	2
8	1	1
39	24	39

[v]

0	0	3
20	5	2
40	1	1
195	24	39

[vi]

0	0	3
0	5	2
36	1	1
99	24	39

[vii]

Stage [i] depicts the setting of the three equations, which in modern notation are as follows:

$$3x + 2y + z = 39$$
$$2x + 3y + z = 34$$
$$x + 2y + 3z = 26$$

Let the right, central and left columns be C_1, C_2 and C_3 respectively. Stage [ii] shows C_2 is multiplied by 3, which is the number in the first row of C_1. Stage [iii] shows C_2 subtracts C_1 successively till its first row is blank. This procedure is called *zhi chu* 直除 which we have translated as "direct subtractions". Stage [iv] shows C_3 is multiplied by 3, which is the number in the first row of C_1. Stage [v] shows C_3 subtracts C_1 successively till its first row is blank. (This case involves only one subtraction.) Stage [vi] shows C_3 is multiplied by 5, which is the number in the second row of C_2. Stage [vii] shows C_3 subtracts C_2 till its second row is blank.

So far, the method is an extension of the method for finding two unknowns, and the rest of the method would also have been similar if the solutions had been integral. Since the solutions involved fractions, some slight deviations from the main method were made to avoid dealing with fractional terms at too early a stage. The remaining steps of the working (stages viii to xi) may be summarised as follows:

[viii] $24 \times 36 - 99 \times 1 = 765$.

[ix] $765 \div 5 = 153$.

[x] $39 \times 36 - 99 \times 1 - 153 \times 2 = 999$.

[xi] $999 \div 3 = 333$.

[xii]   One bundle of low grade cereal $= 99 \div 36 = 2\frac{3}{4}$.

One bundle of medium grade cereal $= 153 \div 36 = 4\frac{1}{4}$.

One bundle of top grade cereal $= 333 \div 36 = 9\frac{1}{4}$.

Some parts of the method are expressed very briefly and our interpretation has been aided by Liu Hui's commentary. (For a translation of Liu Hui's commentary, see [Lam & Ang 1987, pp. 228–234].)

The method given by Sun Zi to solve a set of two equations and that in *Jiu zhang suanshu* to solve a set of three equations reveal that the *fang cheng* method is a general one which can be applied to solve a system of simultaneous equations in any number of unknowns. The Chinese invented a notation when they expressed equations by tabulating numbers in columns. They freed the expression of equations from the entanglement of rhetorics through this notation. The notation is an array of numbers forming a matrix, and the procedure for solving is a process of elimination which results in the remaining non-zero numbers forming a triangle. The method is remarkably modern as it can still be found in our text books; it is usually called "the method of the triangular form".[1] (See, for instance, [Cohn 1958, pp. 19–23].)

The invention of the matrix notation represents an early significant breakthrough in man's struggle to think mathematically in terms of symbols. It is one of those marvellous notations that generates and facilitates

---

[1] The matrix in the present method is the transpose of the matrix in the *fang cheng* method. This difference is probably due to the different ways of writing in the two cultures. The Chinese characters are written in vertical lines from right to left, and our equations are in horizontal lines from left to right.

mathematical thinking. In his discussion on the theory of matrices which was developed in Europe in the 19th century, E. T. Bell [1940, p. 205] made the following apt remarks on the notation: "The invention of matrices illustrates once more the power and suggestiveness of a well-devised notation; it also exemplifies the fact, which some mathematicians are reluctant to admit, that a trivial notational device may be the germ of a vast theory having innumerable applications." The theory of determinants preceded that of matrices, and mathematicians such as A. Cayley claimed that their notion of a matrix was derived either from that of a determinant or as a convenient mode of expressing a pair of linear equations [Bell 1940, p. 205].

The ancient Chinese initiated the notation by using it to express a pair of linear equations, and through it developed the *fang cheng* method of solving simultaneous linear equations. In the 17th century, Seki Kowa of Japan was one of the pioneers of determinants, and his work was influenced by his knowledge of the *fang cheng* method. Mikami [1914, p. 29] made the following observations: "We feel we are pressed to believe in the origin of the Japanese theory of determinants as assuredly derived from the method of solution of the linear simultaneous equations, which the Japanese have learned from their Chinese masters".

In applying the *fang cheng* method to a variety of problems, it would sometimes be inevitable that some numbers in a column would be what we call "negative". This occurrence led to the concept of a new class of numbers, and the need to differentiate between the original class of positive numbers and the new class of negative numbers. According to Liu Hui [Qian ed. 1963, p. 225], the representation of these concepts in rod numerals was as follows: If different coloured rods were used, then red ones represented *zheng* 正 (positive) and black ones represented *fu* 負 (negative). If the rods were of one colour, the numeral representing *fu* (negative) had an extra rod placed diagonally across its last non-zero digit. (For instance, –642 would appear as 丅 ☰ 𠂆.)

It is therefore not surprising that when the author of *Jiu zhang suanshu* discussed the *fang cheng* method, he had to discuss the relationship between positive and negative numbers. This comes under the heading *zheng fu shu* 正負術 (positive-negative rules) followed by four extremely brief statements relating to *zheng* 正 (positive) and *fu* 負 (negative) [Qian ed. 1963, pp. 225–226]. The four statements were written in a stanza pattern, which expressed two sets of rules, one for the subtraction of numbers and the other for their addition. We give below a translation of the four statements. We have added below each statement, our interpretation expressed in terms of modern mathematical notations; a and b are any two numbers, sgn a is sign of a and $|a|$ is the absolute value of a.

Rule 1: Subtraction of numbers

When names are the same subtract mutually; when names are different, add mutually.

When sgn a = sgn b, $a - b = (|a| - |b|)$ sgn a.

When sgn a = $-$ sgn b, $a - b = (|a| + |b|)$ sgn a.

Positive [from] nothing becomes negative, negative [from] nothing becomes positive.

$0 - b = - |b|$ sgn b.

Rule 2: Addition of numbers

When names are different subtract mutually; when names are the same, add mutually.

When sgn a = $-$ sgn b, $a + b = (|a| - |b|)$ sgn a.

When sgn a = sgn b, $a + b = (|a| + |b|)$ sgn a.

Positive [and] nothing becomes positive, negative [and] nothing becomes negative.

$0 + b = |b|$ sgn b.

(For a more detailed discussion of the positive-negative rules, see [Lam & Ang 1987, pp. 235–241].)

How did the Chinese invent the matrix notation? All we can say is that it was an evolution of a fundamental tradition which used positions on the calculating board to represent concepts or things. This property existed in the rod numerals themselves since the positions of the digits of each rod numeral represented the digits' ranks. A fraction was expressed in a concise notation because its concept was embedded in the positions the numerals occupied. Problems were solved by methods which placed numerals in certain positions on the board, such as in the *ying bu zu* method or in those involving proportions and quadratic equations. The matrix notation and the subsequent *fang cheng* method manifested a shift in mathematical thinking towards generalization and abstraction; the outcome was a significant achievement in the removal of linguistic barriers which had impeded the solution of simultaneous linear equations.

The invention of solving a set of simultaneous equations led to a further discovery in the concept of negative numbers, and the subsequent formulation of rules connecting positive and negative numbers. It may safely be said that the ancient Chinese were the earliest to make a mathematical study of negative numbers. In India, Brahmagupta (628) wrote down rules for the addition, subtraction, multiplication and division of positive and negative numbers [Colebrooke 1817, p. 339], and Bhaskara (1150) used a dot to distinguish negative numbers. It was probably around the 15th century that Europeans gradually accepted negative numbers in their own right.

7.8   Miscellaneous problems

We now discuss the remaining problems of *Sun Zi suanjing*. There is one problem in Chapter 2 which has not been mentioned. Prob. 27, which is the same as Ch. 3, Prob. 4 of *Jiu zhang suanshu*, states:

> Now there is a girl who weaves skilfully. Each day she doubles the amount of weaving [done on the previous day]. In 5 days she has woven 5 *chi*. Find the amount she weaves each day.
>
> Answere: First day $1\frac{19}{31}$ *cun*, next day $3\frac{7}{31}$ *cun*, next day $6\frac{14}{31}$ *cun*, next day 1 *chi* $2\frac{28}{31}$ *cun*, next day 2 *chi* $5\frac{25}{31}$ *cun*.
>
> Method: Put down the proportional parts (*lie cui* 列衰) and add them to obtain 31 which becomes the divisor (*fa*). Multiply [the numerals] before they were added up (*wei bing zhe* 未并者) by 5 *chi*, and let each [product] be the dividend (*shi*). Divide by the divisor (*ru fa er yi* 如法而一) to get the answer.

The proportional parts for 5 days would have been displayed on the counting board (see [i]). The addition of the proportional parts gives 31 [ii] which is the *fa* or divisor. Next each proportional part is multiplied by 5

[iii] and then divided by 31 to obtain the answers which are given in *cun* or in *chi* and *cun* [iv].

[i]	[ii]	[iii]

[iv]

Ch. 3, Prob. 4 is often quoted because of its subject matter on Buddhism. It is a simple multiplication problem. Another problem involving straightforward multiplication is Ch. 3, Prob. 13. The problem before this, Prob. 12, has a short cut method. The problem states:

> Now there are 2,374 *hu* of millet. If every *hu* is increased by 3 *sheng*, find the total amount of millet.

We are asked to multiply directly by 1 *hu* 3 *sheng* to obtain the answer.

Ch. 2, Probs. 13 & 14 and Ch. 3, Probs. 6 & 7 have remainders after division. In the answer, the integral quotient and the remainder termed *ji* 奇 are given; *ji* means "an odd lot". In Ch. 2, Prob. 23, the remainder is called *bu jin* 不盡. Other problems (Ch. 2, Probs. 6, 7, 10, 11 & 27, Ch. 3, Prob. 14) express their solutions with the aid of the common fraction.

Ch. 3, Prob. 24 can be said to be involved with a series in arithmetical progression. The problem is as follows:

> Now there is a square bundle of objects whose outer circumference has 32 pieces. Find the total number of objects.
> Answer: 81 pieces.

Method: Put down [32] in two positions. Subtract 8 from the numeral in the upper position and add the remainder to the numeral in the lower position. Continue [in this manner] till there is no remainder, and then add 1 [to the numeral in the lower position] to get the answer.

The stages of computation according to the above description are as follows:

The above steps transcribed into Hindu-Arabic numerals are as follows:

32	24	16	8	0
32	56	72	80	81

We have provided a diagram, Fig. 7.1, so that the method is self-explanatory.

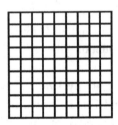

Fig. 7.1

The next problem (Prob. 25) concerns the congruence of two right-angled triangles such that the ratio of their corresponding sides is constant.

Now there is a pole whose length is not known. When its shadow is measured, 1 *zhang* 5 *chi* is obtained. When a model staff of length 1 *chi* 5 *cun* is separately erected, its shadow is measured as 5 *cun*.

Find the length of the pole.

Answer: 4 *zhang* 5 *chi*.

Method: Put down the length of the shadow of the pole, 1 *zhang* 5 *chi*. Multiply by the length of the model staff, 1 *chi* 5 *cun*, and [the product] is raised to tens (*shang shi zhi*) to obtain 22 *zhang* 5 *chi*. Divide by the shadow of the staff, 5 *cun*, to get the answer.

Here we have the multiplication and division of mixed measures of length, which we have already discussed (Sect. 6.2).

Ch. 3, Probs. 27 & 31 are involved with the computation of the number of animals and birds, given the number of heads and legs. The problems and their methods are given below.

Now there are six-headed four-legged animals and four-headed two-legged birds [put together]. [A count] above gives 76 heads and [a count] below gives 46 legs. Find the number of animals and birds.

Answer: 8 animals, 7 birds.

Method: Double the number of legs and subtract from this the number of heads. Halve the remainder to get the number of animals. Multiply the number of animals by 4 and subtract [the product] from the number of legs. Halve the remainder to get the number of birds.

Now there are pheasants and rabbits in the same cage. The top [of the cage] has 35 heads and the bottom has 94 legs. Find the number of pheasants and rabbits.

Answer: 23 pheasants, 12 rabbits.

Method: Put down 35 heads in the upper position and 94 legs in the lower position. Halve the number of legs to obtain 47. Perform repeated [subtractions] by taking away the smaller from the larger [as follows]: the upper 3 is subtracted from the lower 4 and the upper 5 is subtracted from the lower 7; the lower 1 is subtracted from the upper 3 and the lower 2 from the upper 5. The answers are thus obtained.

Another method: Put down the number of heads in the upper position and the number of legs in the lower position. Halve the number of legs and subtract the number of heads, subtract [this result] from the number of heads to get the answers.

The methods are self-explanatory. In the second problem, there is a description for working with rod numerals which we construct below. The description gives a further insight into different ways that subtraction can be performed.

$$\equiv \text{IIII} \qquad \equiv \text{IIII} \qquad \equiv \text{IIII} \qquad = \text{III}$$

$$\underline{\underline{\perp}} \text{IIII} \qquad \equiv \text{T} \qquad - \text{II} \qquad - \text{II}$$

Translated into Hindu-Arabic numerals the above is as follows:

35	35	35	23
94	47	12	12

The alternative method appears to have a similar procedure.

Ch. 3, Prob. 34 is involved with multiples of 9.

> Now there is sighted 9 embankments outside; each embankment has 9 trees; each tree has 9 branches; each branch has 9 nests; each nest has 9 birds; each bird has 9 young birds; each young bird has 9 feathers; each feather has 9 colours. Find the quantity of each.
> Answer: 81 trees, 729 branches, 6,561 nests, 59,049 birds, 531,441 young birds, 4,782,969 feathers, 43,046,721 colours.
> Method: Put down 9 embankments, multiply by 9 to obtain the number of trees. Next multiply by 9 to obtain the number of branches. Next multiply by 9 to obtain the number of nests. Next multiply by 9 to obtain the number of birds. Next multiply by 9 to obtain the number of young birds. Next multiply by 9 to obtain the number of feathers. Next multiply by 9 to obtain the number of colours.

The next problem (Prob. 35) is on combination.

> Now there are 3 sisters. The eldest returns once every 5 days, the second returns once every 4 days and the youngest returns once every 3 days. Find the number of days before the 3 sisters meet together.

Answer: 60 days.

Method: Put down on the right 5 days for the eldest sister, 4 days for the second sister and 3 days for the youngest sister. For each numeral, arrange 1 counting rod on the left. [By performing] *wei cheng* (cross-multiplication), the number of times each sister returns is obtained. The eldest returns 12 times, the second 15 times and the youngest 20 times. Next multiply each by the [corresponding] number of days to obtain the answer.

On the computing board, the display of rod numerals is as follows:

Through the process of *wei cheng* (cross-multiplication), 12 is obtained by multiplying 1 (in the first row), 4 and 3; 15 is obtained by multiplying 5, 1 (in the second row) and 3; 20 is obtained by multiplying 5, 4 and 1 (in the third row). The next display on the board would probably look like this:

Lastly the pair of numerals in each row is multiplied to arrive at the same answer.

The last problem of *Sun Zi suanjing* is on the determination of the sex of an unborn child, given the age of the pregnant woman and the gestation period. Qian [ed. 1963, p. 276] described the problem as "utterly absurd". It is stated as follows:

Now there is a pregnant woman whose age is 29. If the gestation period is 9 months, determine the sex of the unborn child.
Answer: Male.
Method: Put down 49, add the gestation period and subtract the age. From the remainder take away 1 representing the heaven, 2 the earth, 3 the man, 4 the four seasons, 5 the five phases, 6 the six pitch-pipes, 7 the seven stars [of the Dipper], 8 the eight winds and 9 the nine divisions [of China under Yu the Great]. If the remainder is odd, [the sex] is male and if the remainder is even, [the sex] is female.

7.9 Listing the early mathematical methods

To write about the mathematical methods that resulted from the use of the rod numeral system is in fact to undertake the enormous task of writing the greater part of the history of traditional mathematics in China. Here we shall only attempt to compile a brief list of the main achievements which were attained through the rod numeral system. Similar contributions were to be developed at later periods of time through the Hindu-Arabic numeral system.

1. The operations of addition, subtraction, multiplication and division on all numbers, however large. (See Sect. 3.)
2. The notation to express the concept of a common fraction. The use of this notation enabled the addition, subtraction, multiplication and division of fractions, and the processing of other rules related to fractions. (See Sect. 4.)
3. The method of finding the greatest common divisor of two numbers (see pp. 81–82).
4. The representation of a decimal fraction as a numeral. (See Sect. 6.5.) With this representation, the same operations of addition, subtraction, multiplication and division which were applicable to integers were applied to numbers with decimals.
5. The Rule of Three known as *jin you*. (See Sect. 7.2.) This most probably originated from an exchange of goods which had prescribed values. Officials and traders would bring out their rods to reckon the amounts.

6. Methods of solution for partnership and sharing. (See Sect. 7.4.) The positions occupied by the rod numerals on the board represented certain meanings.

7. The *ying bu zu* (surplus and deficit) method (see p. 142). This and its variations solved problems which were involved with surpluses and/ or deficits. Two numerals occupied the upper row and two other numerals were placed below them. The positions of the numerals on the board not only had implied meanings, they were intended for the method of cross-multiplication.

8. The Rule of False Position. (See Sect. 7.6.) This was the *ying bu zu* method, but the rule was applied to problems which did not state explicitly the values and the corresponding surpluses and/or deficits. In such an application, the method presupposed a certain value that entailed a surplus and another value that entailed a deficit. It was a popular method not only in China, but in India and Europe. It was useful at a time when man had not yet attained the level to be able to conceive simple equations in terms of written symbols.

9. The *fang cheng* method. (See Sect. 7.7.) This method solved a set of simultaneous equations through the notational representation of numerals in the form of a matrix.

10. The concept, notation and application of negative numbers (see pp. 152–153). Rules were formulated for the addition and subtraction of positive and negative numbers.

11. The extraction of the square root. (See Sect. 5.) The method was extended to extract the cube root of a number and, later on, roots of higher order.

12. The concept, notation and solution of polynomial equations. (See Sect. 5.4.) These were evolved from the root extraction method.

13. The notation and solution of a set of simultaneous polynomial equations of varying degrees up to four unknowns. The methods were discussed in Zhu Shijie' s 朱世傑 *Si yuan yu jian* 四元玉鑒. (See [Hoe 1977, pp. 91–247].)

14. The Pascal Triangle displayed by Yang Hui and Zhu Shijie. Yang stated that it was derived from an older work written by Jia Xian 賈憲 of the 11th century. (See [Lam 1969a, pp. 82–84].)

15. The method of second order differences interpolation first used by the astronomer Liu Zhuo 劉焯 (544-610 AD). The method of computing with the unequal interval second differences interpolation formula was recorded in *Da yan li* 大衍曆, which was complied by Yi Xing 一行 in 727. (See [Li & Du 1987, pp. 88–92; Ang 1979, pp. 326–340].)

16. The method of third order differences interpolation adopted by Wang Xun 王恂 (1235–1281), Guo Shoujing 郭守敬 (1231–1361) and others. (See [Li & Du 1987, pp. 151–155].)

17. The technique of stacking piles. A simple problem is illustrated in *Sun Zi suanjing* Ch. 3, Prob. 24 (see p. 155). Shen Kuo 沈括 (1032–1095) initiated the investigation into higher order differences series. He was followed by Yang Hui [Lam 1977, pp. 18 & 232] and Zhu Shijie [Hoe 1977, pp. 300–321].

18. The well known remainder problem in *Sun Zi suanjing* which was discussed in Sect. 7.5.

19. In the 13th century, Qin Jiushao was able to give a complete method of solution to the remainder problem, even for the case where the moduli were not relatively prime. (See [Libbrecht 1973, pp. 328–366; Li & Yuan 1983].)

Socioeconomic Aspects in Sun Zi's China

8.1 The period

As a rudimentary book in mathematics, *Sun Zi suanjing* was written for the widespread use of probably both functionaries and the common people. The text reveals some interesting socioeconomic aspects of life in ancient China which we shall discuss in this section.

As we have mentioned earlier (Sect. 1.2), the dating of the book is uncertain being somewhere between the third century and the beginning of the fifth century AD. This was a period that saw China being divided into Three Kingdoms (220–264) and a subsequent reunification under Western Jin (265–317) followed by Eastern Jin (317–420). Throughout the third century short-lived and weak Chinese dynasties managed to hold North China, but from 311 onwards the fortunes of the Chinese began to decline. The *xiongnu* 匈奴 or Huns captured the city of Luoyang in that year, and five years later Chang'an also fell. This marked the end of Chinese control of the north which was to last for almost three hundred years. The fall of the two centres of Chinese civilization led to an exodus of officials, literati and others from the north to the south. Fleeting glimpses of such scenarios could be seen from some of the practical problems in *Sun Zi suanjing*.

Prob. 33 of Chapter 3 gives the distance between Chang'an and Luoyang as 900 *li*. The period of unrest and disunity could have ruined much of the splendour of the two cities, so that when the protracted wars ended and

China was reunified under the Sui dynasty (581–618), both Chang' an, the western capital, and Luoyang, the eastern capital, had to be resited and reconstructed. Chang' an was rebuilt in the year 582 at a site 20 *li* southeast of the old city, and in 605, Luoyang was shifted 18 *li* west of the old site [*Jiu Tang shu* Ch. 38, pp. 1,394 & 1,420]. According to *Jiu Tang shu* [Ch. 38, p. 1,422], the distance between Chang' an and Luoyang at that time was 850 *li*.

On the political side, Sun Zi' s problems reveal to some extent the unstable and vulnerable conditions of the time, which warranted the conscription of soldiers. Ch. 3, Probs. 2 & 7 give us the rates of conscription of soldiers to be one out of every 37 men and one out of 25.

8.2 Buddhism

Since the introduction of Buddhism into China in the first century AD, Luoyang had become a Buddhist centre, particularly for the translation of Buddhist texts [Ch' en 1964, p. 43]. By the third and fourth centuries, the impetus for the growth and spread of Buddhism was tremendous. Translation of Buddhist works was actively carried out by both Indian and Chinese monks. It has been estimated that "the volume of translated works steadily increased, from an average of 2.5 works translated per year, in the period up to 220 AD, to 9.4 works per year in the period 265–317" [Wright 1959, p. 40]. The book on Buddhism that Sun Zi mentioned in Ch. 3, Prob. 4 could be a translated work. It was probably a common sutra, and since it was short with only 1,827 characters, it could be copied quickly for the purpose of transmission.[1]

8.3 *Wei qi* and the nobility

Apart from the conducive climate favourable for the study of Buddhism, we have the picture of people playing a leisure game of *wei qi* 圍棋 (encirclement chess). This was a favourite intellectual and social pastime

---

[1] Yan [1937, p. 313] even attempted to identify the translated text to be *Fo shuo a mi tuo jing* 佛說阿彌陀經 by Zhi Qian 支謙, a national of Yue Zhi 月支 (Scythia).

among court officials and scholars. Ch. 3, Prob. 5 indicated that the game was played on a board of 361 squares. The game was played with chips of black and white by two players, the object of each player being to surround the pieces of the opponent so that he would be unable to move. The game was made popular during Sun Zi's time, and continued to be a favourite pastime inside and outside official circles in subsequent centuries.[2]

In the sphere of the ruling nobility, a problem from *Sun Zi suanjing* (Ch. 2, Prob. 25) indicates the ranking of noblemen and the ratio of their respective entitlements during the author's time. The practice of conferring titles on members of the imperial family and distinguished personalities could be traced to as far back as the Zhou dynasty in the pre-Christian era. There were twelve titles of nobility, comprising three varieties of princes (*wang* 王), five of dukes (*gong* 公), and one each of marquis (*hou* 侯), earl (*bo* 伯), viscount (*zi* 子) and baron (*nan* 男). During Sun Zi's time, the hierarchy was reduced to the last five ranks. In the Northern Wei dynasty (386–534), it was further reduced to four classes, namely, prince, duke, marquis and viscount [*Wei shu* Ch. 113, p. 2,973]. This change was reflected in a problem of *Zhang Quijian suanjing* [Qian ed. 1963, p. 342].

In Sun Zi's problem, we are informed of the ratio of entitlement for the different ranks in the distribution of goods. The nobility also enjoyed other privileges such as the distribution of land and the allocation of households under their control. The account given in Sun Zi's problem is in line with that mentioned in *Jin shu* [Ch. 14A, pp. 414–415].

8.4   Taxation

We can view the taxation system during Sun Zi's time through Ch. 3, Probs. 1, 6, 9 & 10. Wu Xianqing 武仙卿 postulates that the household taxation system was inaugurated by the Wei kingdom in the year 225 at the rate of two *pi* of silk and two *jin* of floss per household.[3]   (See [Yan 1937, p. 312].)

---

[2] An expert *wei qi* player could be rewarded with certain privileges. For example, Wang Shuwen 王叔文 of the 8th century was sent to Chang'an because of his skill in the game. He was subsequently appointed *shidu* 侍讀 (reader-in-waiting) to the heir apparent in the imperial household [*Jiu Tang shu* Ch. 135, pp. 3, 733].

[3] Yang Lien-sheng [1961, p. 179, fn. 104] states that the household tax system was probably first instituted in the year 204 or earlier.

The taxation policy after the defeat of the Wu kingdom (222–280) by the Western Jin dynasty was recorded in *Jin shu* [Ch. 26, p. 790] as follows: "Each household headed by a regular male adult paid annually three *pi* of silk and three *jin* of floss. A household headed by a female or a secondary male adult paid one half of the taxes. The levies were sometimes reduced to two-thirds for prefectures along the boundary and to one-third for the farthest ones" [Yang 1961, p. 179]. In Ch. 3, Probs. 9 & 10, the rate of payment in terms of floss was given as 2.5 *jin*. The total of 36,454 households mentioned in the problems could have been the actual figure in the region where Sun Zi lived. There is a possibility that he might have taken an average rate of payment of 2.5 *jin* per household, after taking into consideration households along the boundary and those headed by a female or a secondary male adult.

Yang Lien-sheng [1961, p. 129] points out that during the period of the Three Kingdoms and the Jin dynasty, "the government endeavored to induce people to settle down through the encouragement of agriculture, the creation and repair of water works, and the establishment of civil and military agricultural colonies". A male adult received from the government a certain amount of arable land for ploughing. This allotment was to be returned to the government when the recipient was old or dead. A female received about half the allotment of the male. Both male and female were further permitted to own a certain amount of land as household property. According to *Jin shu* [Ch. 26, p. 790], a male was alloted 50 *mou* of land by the government and a female 20 *mou*; a male was permitted to own 70 *mou* and a female 30 *mou*.

A land tax at the rate of 4 *sheng* of grains per *mou* was levied towards the end of the Han dynasty. Yang Lien-sheng [1961, p. 142] points out that it is not clear whether this rate was applied to all grades of cultivated land or was only an average. Basing his data from the *Jin shu*, Yang goes on to say that a tax of 3 *sheng* per *mou* was levied in the year 330 when the fields of the people were surveyed for the first time. The rate was subsequently reduced to 2 *sheng* per *mou* in 362. In 377, the system of levying taxes in proportion to land measurements was abolished and a flat rate of 3 *hu* per person was instituted instead. Six years later the rate was increased to 5 *hu* per person.

Ch. 3, Probs. 1 & 6 not only provide information about taxes in kind paid by the respective households, they also show the expensive cost in transportation. A cart at that time could only carry 50 *hu* of grains. For every 1,000 *hu* of grains sent to the national granary, 200 *hu* would be deducted as expenses for transportation which was to be borne by the taxpayers. For this reason, Prob. 1 says that each household had to contribute its share of transport expenses.

It can be gauged from the contribution of land tax paid by the households that the yield from the land was quite considerable. The amount that went into the national granaries, too, must be enormous. This is substantiated by a record in *Jin shu* [Ch. 47, p. 1,321] which says that the Wei rulers had no need to worry about the income from the land tax, because the supply was assured by the large numbers of military and civilian tenants who sent 50 to 60 percent of their land produce as rent to the government. This rate was observed through most of the Jin period. There are four problems in *Sun Zi suanjing* (Ch. 2, Probs 10, 11 & 12; Ch. 3, Prob. 3) which mentioned granaries. The granaries of the first two problems were extremely large: they could hold 151,474 *hu* $7\frac{11}{27}$ *sheng* and 53,666 *hu* 6 *dou* $6\frac{2}{3}$ *sheng* respectively. They were probably national granaries. The granaries of the last two problems, which had capacities of 2,700 *hu* and 100 *hu* respectively, appeared to be household granaries. The surplus of grains was sometimes stored in a granary for as long as nine years as stated in Ch. 3, Prob. 13.

When the supply of grains exceeded demand, the price would naturally drop. An illustration of such a situation is found in *Jin shu* [Ch. 26, p. 786]: "When the Jin received the mandate of Heaven [around 265], Emperor Wu wished to pacify and unify the territory south of Yangzi River. At that time grains were cheap, but cloth and silk were expensive." The account goes on to say that the Emperor instituted a system of exchanging cloth and silk for cheap grains which were then stored away, the object being to maintain the grains at a stable fair price. This measure was first introduced in the state of Wei during the Warring States period and was practised in the Han dynasty. (See [Yang 1961, p. 170].)

8.5   The barter trade

Barter trading was freely practised during Sun Zi's time. One reason for this practice was probably due to the abundant supply of grains, and the other reason was the shrinkage in the use of metallic money called *qian*. (See [Yang 1961, p. 129] on Quan Hansheng's 全漢昇 research findings.)

During the third to the fifth century, copper coins were frequently reported to be out of circulation. It was therefore quite common for most levies during Sun Zi's time to be collected in kind. This practice was especially true in northern China. Prices of goods were often quoted in terms of the number of pieces of silk or cloth. Due to the high value of silk, the latter had become the target of robbery. Sun Zi drew attention to this subject in the last problem of Chapter 2.

Ch. 3, Prob. 14 informs us the interest rate for the loan of silk was approximately 28 percent per annum. In Ch. 3, Probs. 11 & 16, Sun Zi gave a comparison of the worth of millet with beans and silk respectively. It is also interesting to note the value of *qian* in those days: 1 *jin* of gold is worth 100,000 *qian* (Ch. 3, Prob. 20) and 1 *pi* of brocade is worth 18,000 *qian* (Ch. 3, Prob. 21).

Did the Hindu-Arabic Numeral System Have Its Origins in the Rod Numeral System?

9.1 The background leading to such an investigation

In the preceding sections, we discussed the concept of the Chinese number system, the structure of their numerals, and how they were employed in the various operations of arithmetic. The problems of *Sun Zi suanjing* illustrated the development of arithmetic and showed how this was extended to the field of algebra. Generally the development of arithmetic and algebra in ancient China was similar to that which occurred in Europe from the 12th century onwards, after the adoption of the Hindu-Arabic numeral system.[1] (See Sect. 7.9.) In other words, the main concepts which provided the foundation of our arithmetic and algebra were already known to the ancient Chinese. The similarity is all the more remarkable when we take into consideration the time difference of several centuries between the two developments, and the fact that one development was fostered through a computation rod numeral system while the other was through the written Hindu-Arabic numeral system.

Was this similarity a mere coincidence? Why did the ancient Greeks, who produced geniuses like Euclid and Archimedes and made significant advances in geometry, fail to develop arithmetic and algebra along such

---

[1] One of the earliest and most important European work is *Liber Abbaci* (1202) by Leonardo Pisano. (See [Boncompagni 1857].)

lines? Why did some civilizations succeed at arithmetic, while others failed, at what might appear a relatively simple task?

The answer is perhaps, that simple as it is today to add, subtract, multiply and divide, such arithmetic, which is easily accessible to everyone, was not an easy discipline to invent.

Arithmetic operations are inseparable from the numeral system from which they arise. The development of a numeral system, is by any standard, a significant conceptual feat. However, as will be elaborated later (Sect. 9.4), not all numeral systems can support anything more than rudimentary arithmetic. To develop a numeral system which can lead to the higher reaches of mathematics is therefore a feat of considerable originality — such originality, that most ancient civilizations did not possess it.

Long before the worldwide use of the Hindu-Arabic numeral system, various ancient civilizations evolved their own numeral systems. If we wish to understand the development of their arithmetic, we have to understand the concepts of their number systems, their transcription processes into numerals and other devices used for reckoning. At the same time, we have to constantly guard against using our own preconceived notions, derived from using the Hindu-Arabic numeral system, to understand the other systems. We need to have an understanding from within the ancient system. The probe into the development of arithmetic in China, from this perspective, has led one of us to point out that China was the earliest civilization to possess the concept of the Hindu-Arabic numeral system [Lam 1986, 1987], and to advance the thesis that the Hindu-Arabic numeral system had its origins in the Chinese rod numeral system [Lam 1988]. We summarise in this Section the arguments that have been put forward.

9.2   The rod numeral system and the Hindu-Arabic numeral system have the same concept

In Sect. 2.2, we analysed the structure of the Chinese written number system and showed that it has the unique feature of representing both number words and number symbols. In other words, the concise ideograms not only express the meaning of a number in a written text, but can also function as numerals. This number system requires ideograms only for the first nine numbers and numbers in powers of ten. The numbers are built in gradations in units, tens, hundreds, thousands, and so on, and the ideographic characters representing numbers in powers of ten also represent numerical ranks. Any other number (or numeral) is made up of characters from the first nine numbers (一二三四五六七八九), which we call digits, and characters representing numerical ranks. For example, in 九千六百三十四 (9,634), 九, 六, 三 and 四 are the digits and 千, 百 and 十 are the numerical ranks. The structure of this written number system is the only one of its kind in world history.

A number is thus composed of quantities expressed in terms of numbers in powers of ten. A study of the evolution of the forms in the written number from the oracle-bone writings to the present day ideograms revealed that the concept of the number system remained basically unchanged.

The rod numeral system (Sect. 2.3) differed from the written number system in that it dispensed with the need for signs to represent ranks: instead it used the position occupied by each digit of a numeral to represent the rank of the digit. The rod numeral system thus used a place value notation with ten as base and needed only nine signs which by themselves, represented the first nine numbers. The digits of a rod numeral were arranged in a horizontal row from left to right in descending order of their ranks, beginning with the highest. For instance, *jiu qian liu bai san shi si* 九千六百三十四 (9,634) would appear as ⊥ T ≡ |||| , with the concepts of *qian* 千 (thousands), *bai* 百 (hundreds) and *shi* 十 (tens) transferred to the places occupied by 9, 6 and 3 respectively. When a number had no digit of a particular rank, the position representing that rank on the board was left vacant; for example, *jiu qian liu bai si* 九千六百四 (9,604) would appear as ⊥ T |||| .

The Hindu-Arabic numeral system is constructed on the same concept as the rod numeral system: it uses a place value notation with ten as base, and the nine signs needed are those which represent the first nine numerals. A number such as nine thousand six hundred and thirty four would be written as 9634, where the rank of each digit is notated by the position it occupies. The digits of a numeral, like those of a rod numeral, are arranged horizontally from left to right in descending order of ranks, commencing with the highest. How is a numeral written when it has no digit of a particular rank? In the present notation, a tenth sign, which is the zero symbol, is written on the position representing that rank, so that nine thousand six hundred and four would be 9604. However the earliest form of the Hindu-Arabic numerals did not use this symbol, but like the rod numerals, had a blank space in its stead, so that the above number would appear as 96 4. This space was called *sunya* in India and *sifr* in Islam, and both words meant empty. In the rod numerals, the Chinese called the vacant space *kong* 空 which also meant empty.

9.3   On the significance of the concept of an invention

The rod numeral system and the early Hindu-Arabic numeral system are thus conceptually identical. Furthermore they also share the same convention of arranging the digits of a numeral in a horizontal row from left to right in the descending order of their ranks. They have another conventional similarity in the case when a numeral does not have a digit of a particular rank; the space representing that rank is left vacant.

In what ways are the two systems different? The symbols for the nine signs are not the same, and one system is expressed through rods while the other is a written one. However we cannot dismiss the theory of a Chinese origin of the Hindu-Arabic numeral system merely on the grounds of these differences. What matters far more for the question of origin is whether the numeral systems are similar in concept, not whether they are expressed in similar symbols. The primary importance of an invention is the idea behind it and not the superficial form that expresses the invention. It is significant to emphasize here the distinction between the evolution of a concept and the evolution of the different forms of expressing the attained concept. In the first case, it takes a genius or somebody close to one to achieve a concept which has lasting implications; in the other case, it can be anyone who has fully understood the concept and is able to suggest better ways to express it. These expressions would naturally be in the form and medium befitting the time and place.

The rod numerals were introduced not later than the Warring States period (475–221 BC) at a time before paper was invented, when writing was fraught with difficulties. From the primitive stage of reckoning with simple numbers through the use of rods made from animal bones or wood, the device evolved into the elegant rod numeral system, which was used in China for well over a thousand years. It was easily accessible to foreigners for a considerable span of time. Would one of them have adopted the idea, but changed the signs to fit a different culture? If he lived in the paper era, would he have upgraded the system into a written one? Any of such possibilities would constitute an attempt to change the superficial mode of the invention.

Given that the rod numeral system is conceptually identical to the Hindu-Arabic numeral system, it is perfectly possible that the latter was derived from the former. To establish this, we need to dispel two alternative theses: (A) that the concept of the Hindu-Arabic system was transmitted to India from a system other than the rod numeral system; (B) that it was an invention of Indian origin.

9.4 No other numeral systems of antiquity share the same concept as the rod numeral system

A study of all the known numeral systems of antiquity reveals that the rod numeral system is the only one which is conceptually identical to the Hindu-Arabic system. Some of these systems, such as the Babylonian cuneiform system and the Mayan system, have the place value feature but their number bases are not decimal. Others such as the Egyptian hieratic system, the Greek alphabetic system and the Roman system do not have the place value notation. Therefore they require a system of notations to represent the numbers 1, 2, ..., 9; 10, 20, ..., 90; 100, 200, ..., 900; and so on, so that more notations have to be remembered as the numbers grow larger. The two systems of ancient Indian numerals from the Kharosthi and Brahmi scripts are also similar in this respect [Datta & Singh 1935, pp. 22, 26]. See Figs 9.1 to 9.7. Even the unique Chinese written numeral system requires new ideograms to express numbers in higher powers of ten.

Thus, the only numeral system that is conceptually identical to the Hindu-Arabic system is the rod numeral system. Thesis A advanced above — that the concept of the Hindu-Arabic system was transmitted to India from a system other than the rod numeral system, cannot be true.

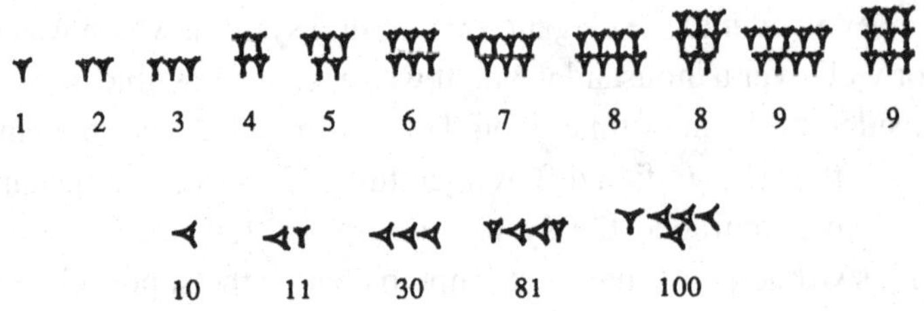

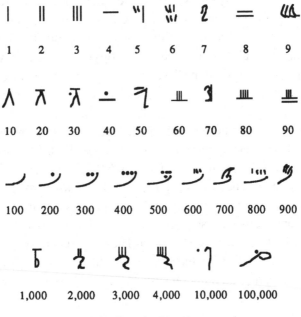

Fig. 9.1   Babylonian cuneiform numerals

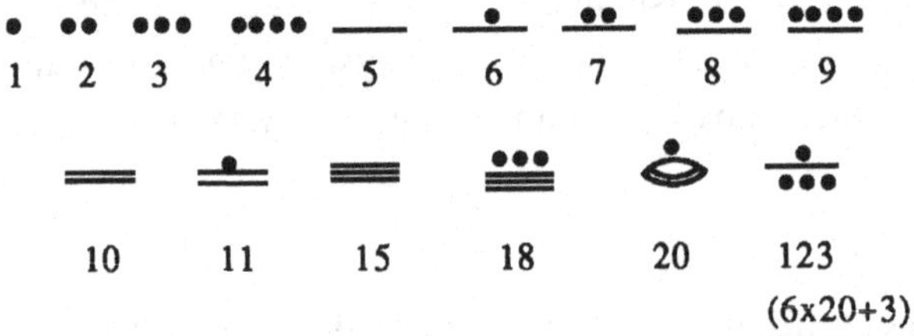

Fig. 9.2   Mayan numerals

Fig. 9.3   Egyptian hieratic numerals

α	β	γ	δ	ε	F	ζ	η	θ
1	2	3	4	5	6	7	8	9

i	κ	λ	μ	ν	ξ	ο	π	Ϙ
10	20	30	40	50	60	70	80	90

ρ	σ	τ	υ	φ	χ	φ	ω	ϡ
100	200	300	400	500	600	700	800	900

Fig. 9.4   Greek alphabet numerals

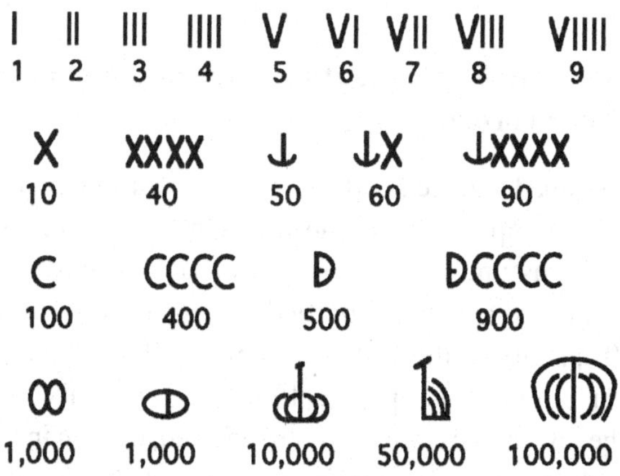

I	II	III	IIII	V	VI	VII	VIII	VIIII
1	2	3	4	5	6	7	8	9

X	XXXX	↓	↓X	↓XXXX
10	40	50	60	90

C	CCCC	Đ	ĐCCCC
100	400	500	900

∞	⊕	⊕	𝕝	(((·)))
1,000	1,000	10,000	50,000	100,000

Fig. 9.5   Roman numerals

I	II	III	X	IX	IIX	XX	?
1	2	3	4	5	6	8	10

ʒ	?ʒʒ	ʒʒʒ	ʒʒʒʒ	人	𝓏II
20	50	60	70	100	200

Fig. 9.6   Kharosthi numerals

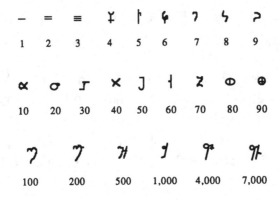

Fig. 9.7 Brahmi numerals

9.5 On the hypothesis that the Hindu-Arabic numeral system has an Indian origin

It is generally assumed that the Hindu-Arabic numeral system has its origins in India. However, up till now, there has not been any satisfactory explanation for this assumption. Scholars who have hypothesized an Indian origin, have admitted that they are baffled as to how the system originated. Dantzig [1930, p. 29] credited the system to "the achievement of the unknown Hindu". Smith & Karpinski [1911, p. 41] had this to say, "Who it was to whom the invention is due, or where he lived, or even in what century, will probably always remain a mystery". Although Menninger [1969, pp. 393–399, 460] assumed an Indian origin, he admitted that the beginnings were obscure. Cajori [1928, p. 46] remarked that the time and place of origin had not been settled.

Cajori [1928, p. 46] also drew attention that there were three investigators, working independently of one another, who had questioned the assumption of an Indian origin. One of the investigators,[2] Kaye [1907, p. 503], drew the following conclusions: "The task I set myself was to show that the current conceptions as to the origin of our modern arithmetical notation have not very secure foundations and that the question is worth reopening.... The character of the Indian scripts; the evidence of inscriptions;

---

[2] The other two were Carra de Vaux and Nicol Bubnov.

the nature of the early notations in use among the Hindus; the nature of their mathematical works; the very custom at the present time among those Hindus who work on purely indigeneous lines point to a foreign origin of the modern notation as probable...".

The Indians themselves do not know how the Hindu-Arabic numeral system originates. Datta & Singh [1935, p. 49] said that the inventor of the system was unknown. There are no early Indian treatises which give a description of the numerals nor are there any which show the fundamental operations of arithmetic associated with the system. According to Datta & Singh [1935, p. 40] the earliest epigraphic evidence of Hindu-Arabic numerals in India was in 595 AD.

The postulation of an Indian origin is based mainly on three reasons: (i) The symbols denoting the nine signs of the Hindu-Arabic numeral system have an Indian origin. (ii) The Arabic texts, which describe the numerals and through which they were transmitted to Europe, call them Indian or Hindu numerals. (iii) The topics on arithmetic and algebra of some Indian mathematical treatises correspond closely to the subjects associated with the Hindu-Arabic numeral system.

These reasons are not sufficient for establishing an Indian origin. The latter two reasons are perfectly consistent with the possibility that the system was transmitted from another country to India, followed by the transmission from India to the Arab countries. For instance, while some of the early topics on arithmetic and algebra associated with the Hindu-Arabic system could be found in Indian treatises, they could often be found in Chinese texts of an earlier date.

As for the first argument regarding the shapes of the Hindu-Arabic numerals, it is generally believed that the nine signs could be traced to the first nine Brahmi numerals. However the Brahmi numeral system is made up of a larger number of signs and lacks the place value concept. (See Fig. 9.7.) It could not have been the conceptual precursor of the Hindu-Arabic system.

The history of the shapes of the Hindu-Arabic numerals themselves has been long and tortuous. (See, for example, [Hill 1915].) It was the advent of printing that largely formalised the shapes that we are now familiar with. According to Smith & Ginsburg [1937, p. 20], the origin of the shapes of most of the numerals was not known — priests simply invented their own letters and numerals. They went on to say, "As travel became easier and as merchants and rulers felt the need for writing, the numerals of the various tribes tended to become more alike." Smith & Karpinski [1911, p. 14] said, "It should be stated definitely at the outset, however, that we are not at all sure that the most ancient forms of the numerals commonly known as Arabic had their origin in India. As will presently be seen, their forms may have been suggested by those used in Egypt, or in Eastern Persia, or in China, or in the plains of Mesopotamia." Cajori [1896, pp. 13–14] listed a number of facts concerning the origin of the shapes of the numerals that required "to be explained and harmonized". He then commented, "Need we marvel that, in attempting to harmonize these apparently incongruous facts, scholars for a long time failed to agree on an explanation of the strange metamorphoses of the numerals or the course of their fleeting footsteps as they migrated from land to land?"

9.6 Evidence of a Chinese origin

In the preceding section, we have shown that there is little evidence to support the theory of an Indian origin. We shall now present evidence from *Sun Zi suanjing*, that supports the thesis that the Hindu-Arabic numeral system originated from the rod numeral system.

The ancient Chinese invented the rod numeral system primarily to perform addition, subtraction, multiplication and division. In the same vein, the Hindu-Arabic numeral system was adopted by countries all over the world because it was by far the most superior means of performing these operations then known. An investigation into the origins of numeral systems should therefore consider the original methods of performing the four fundamental operations of arithmetic.

Sun Zi suanjing gives a detailed description of how multiplication and division were performed with rod numerals. In Sects. 3.3 and 3.4, we have shown that the step by step procedures are identical with the earliest methods of performing multiplication and division using Hindu-Arabic numerals, as described in three Arabic texts. These are: a Latin translation of al-Khwārizmī's work on arithmetic, al-Uqlīdisī's *Kitāb al-Fusūl fī al-Hisāb al-Hindī* and Kūshyār ibn Labbān's *Kitāb fī Usūl Hisāb al-Hind.*

Although the methods of addition and subtraction with rod numerals are not explicitly described in *Sun Zi suanjing*, they can be inferred from the addition and subtraction procedures of the multiplication and division methods. In Sect. 3.5, we have shown that the methods are similar to the earliest procedures using Hindu-Arabic numerals described in the above three Arabic texts.

The ancient Chinese evolved the techniques of adding and subtracting two rod numerals through the intrinsic properties of the numerals themselves. The numerals are placed in two rows one above the other. The procedure for adding and subtracting two rod digits is based on the simple rules of operating with horizontal and vertical rods representing fives and units respectively, or units and fives respectively, depending on the place values of the digits. (See Sect. 3.2.) The process involves either removing rods or placing additional rods in the correct vertical or horizontal position as the case may be. The operation on each pair of corresponding digits begins from left to right, and changes are made to the digits of the numeral in the upper row which finally displays the answer. (See Sect. 3.5.) The operations of multiplication and division also involve the same principles. Like addition and subtraction, they commence from left to right. They require knowledge of the multiplication table, and both procedures involve shifting the multiplier or divisor place by place from left to right. The rules of arithmetic operations thus arose from an accepted usage through performance with rod numerals.

If we allow the possibility of an independent invention of the Hindu-Arabic numeral system, how does one explain that such a system, which is a written one with a set of symbols so different from those of the rod numerals, evolved identical procedures of addition, subtraction, multiplication and division as those of the rod numeral system? When a set of procedures is adopted, this is merely a matter of convention and is inevitably related to the cultural background of the system. An explanation is therefore required as to how the same arithmetic procedures can be adopted by two different cultures separated by a lengthy span of time.

Al-Khwārizmī's book, translated into Latin in the 12th century, was the first on the new system of numeration to reach Europe. There are numerous ways of multiplying and dividing with numerals;[3] the fact that the detailed

---

[3] It is interesting to note Boyer's [1944, p. 162] suggestion on the application of multiplication with some ancient numerals.

steps for these two operations in the two treatises, the Chinese written around 400 and the Arabic about 825, are virtually identical, is a remarkable phenomenon. This is especially so when we take into account how different the two media of computation were.

As explained above, the arithmetic procedures were a natural outgrowth of a rod and board system. On the other hand, some of the procedures, such as operating from left to right, the constant changes in the digits of a particular row, the removal of digits, the advancing of the multiplier or divisor to the left and the shifting of the multiplier or the divisor place by place from left to right, show that they are not conducive to a written system.

That those very procedures have been shown to be the earliest in Islam suggests the conclusion that they were adopted in *toto* from a non-written numeral system. In fact, after the numeral system and arithmetic procedures had been understood and assimilated, numerous other methods of operations started to appear in both Islam and Europe. They were either improvements from the original ones or were constructed quite differently, but in all cases the changes were made for the convenience of a written medium. For instance, take the galley method of division, known also as the scratch method, which was popular in 15th century Europe. This method was the original Chinese-Arabic method of division, but since it was in a written form with a step by step format, the stages where rod digits were either removed or changed on the board were transcribed into cancelled digits on paper. (See, for example, [Smith 1925, pp. 136–138].) The methods of performing the arithmetic operations with which we are now familiar have been the result of an evolution to attain more convenient ways for a written system.

9.7 Other identical conventions

Perhaps the best way to illustrate the improbability of an independent Indian origin of the Hindu-Arabic numeral system would be to consider the various other conventional similarities between the Hindu-Arabic and the rod numeral system. Apart from the identical procedures used in the four basic arithmetic operations discussed above, the two systems also have identical procedures for the arrangement of digits, the method of representing "zero" and the notation for fractions. We shall now touch on each of these in turn.

In a rod numeral, the digits are arranged in a horizontal row from left to right, in descending order of their ranks, beginning with the highest. The same conventional form was adopted in the display of the Hindu-Arabic numerals.

When a rod numeral had no digit of a particular rank, the position representing that rank on the board was left vacant. The empty space, which was called *kong* 空, meaning empty, was a natural process which resulted from the use of the rods. To display a number, the rod symbol for each of its digits had to be slotted into the correct position for its numerical rank; so if there were no digit of a particular rank, the position of that rank would then be left vacant (see p. 49). The earliest Hindu-Arabic numerals followed this convention; the vacant space without a digit was called *sunya* in India and *sifr* in Islam, and both words meant empty. Being a written system with a set of symbols different from the rods, the use of blank spaces to denote the absence of digits created ambiguity in the values of the numerals. Later the empty space was replaced by the zero symbol and in some places by a dot.

The fact that the Chinese had used the rod system for many centuries, obviously showed that they did not have any problem with the *kong* in their rod numerals. The rod system was essentially for computing, and after the computer had completed his calculations and recorded the results in the written numerals, the rods would be cleared away. In this short time, the computer would therefore remember what numerals he had laid on the board. Moreover the rod digits were of two types (see Sect. 2.3.1), which depended on the ranks of their positions, and the numerals would be placed relative to one other. All these factors made the numerals easily discernible. The blank space in the rod numeral structure was thus a natural outgrowth of a system working with rods. On the other hand, it would be rather odd to find a blank space within a written numeral, if the written system were an original invention. A symbol such as a dot or a circle would probably have been a more natural convention to adopt, right from the start.

We now move to the identical notation for fractions. In Sect. 4.1, we have shown how the concept of a common fraction was embodied in the notational rod numerals left behind in the last stage of computation, when two numerals were divided leaving a remainder. Thus, the division of 100 by 6 gives $16\frac{4}{6}$, and this was expressed on the counting board as

$$
\begin{array}{c}
-\ \mathsf{T} \\
\mathsf{IIII} \\
\mathsf{T}
\end{array}
$$

The fraction $\frac{4}{6}$ was notated as $\frac{||||}{\mathsf{T}}$. All fractions on the counting board assumed such a notation: the numeral of the fractional part was placed above the numeral representing the whole. In Sect. 4, we have also shown that it was through the use of this mathematical notation that the ancient Chinese were able to develop the study of fractions to a level very similar to the arithmetic manipulations of fractions that we know today.

The earliest procedures for performing the four fundamental operations of arithmetic in the Hindu-Arabic numeral system were identical with those of the rod numeral system. The identical processes of division led to identical formats in the last stage, and this in turn led to identical forms of expressing the concept of a fraction. Thus, the fraction $\frac{4}{6}$ was also notated in the form $\frac{4}{6}$ for the Hindu-Arabic system, with the numerals written in the symbols of those times. The extra horizontal line was probably introduced in the 12th century [Cajori 1928, p. 269].

We have pointed out that when one set of procedures is adopted, it is merely a matter of convention. When a fraction is expressed in a particular form, this is also only a matter of convention. For instance, the ancient Egyptians and Greeks expressed their fractions very differently; the Egyptians had notations only for unit fractions while the Greeks differentiated the fractions from their alphabetic numerals by using accent marks and other devices (see [Cajori 1928, pp. 26–27; Flegg ed. 1989, pp. 131–143]). Therefore, the fact that the Arabs and Chinese had identical forms of expressing the complicated concept of a fraction, just as they had identical arithmetic procedures and identical forms in the numerals, cannot be dismissed as mere coincidences. Given that the Chinese had evolved all such forms and procedures at a significantly earlier date, this inevitably points to the Chinese origins of the Hindu-Arabic numeral system.

9.8   On the question of transmission

Two main strands of evidence have led us to the conclusion that the Hindu-Arabic numeral system originated from the rod numeral system: first, the rod system is the only numeral system to possess the concept of the Hindu-Arabic system, and second, the earliest fundamental arithmetic procedures using Hindu-Arabic numerals are identical with those that were used with rod numerals. While similar concepts may occur independently in different civilizations at different periods of time, it would be too much of a coincidence for the procedures of these inventions to be identical too.

However we are still faced with the question of how the concept of the Hindu-Arabic system was transmitted from China. This could plausibly have arisen as follows:

Rod numerals were used in China for a very long period of time — a continuous stretch of more than one and a half millennia. They had been in use since the Warring States period (475–221 BC) till their gradual demise in the latter half of the Ming dynasty (1368–1644), when they were replaced by the abacus.

The rod numeral system was commonly used by all kinds of people ranging from astronomers, mathematicians and scientists, to administrations, Buddhist monks, traders and travellers. The rods were carried and used whenever and wherever computation was required. It is well known that during the Tang dynasty (618–907 AD), civil and military officers carried a bagful of rods with them wherever they went.

East of China, the rod numeral system was transmitted in *toto* to Korea and then to Japan around the 6th century. The rods were called *sangi* in Japanese, and the system was practised for a few hundred years in both countries.

It would therefore be no surprise if the Chinese reckoning method reached India and countries further west, given the extensive contacts fostered by the trade routes. It is possible that the rod numeral concept could even have been transmitted to two different places via different routes, in such a way that the local adaptations into written notations would have been radically different.

This is a plausible explanation of the hitherto inexplicable fact that two different versions of the Hindu-Arabic numerals were found in the Arab countries. These were the East Arabic and West Arabic numerals [Menninger 1969, pp. 413–417]. The zero of the East Arabic numerals was represented by a dot and the West Arabic numerals did not have a zero. The latter, also called *gubar* numerals, had dots placed above the digits to indicate their order. If one of the two Arabic numeral systems had sprung from the other, or if both of them had had a common origin in an Indian or an unknown written system, the radical dissimilarities in notations would be hard to explain. For all three systems would have had written notations, and it would be difficult to imagine how a written notation could undergo such marked transformation in its transmission to two regions in such proximity.

However if both Arabic systems had been transmitted via different routes from the Chinese rod system, then the mystery disappears. For the Chinese system is not in written notation, and when two regions independently transcribe the rod system into written numerals, very different results can be expected.

It is interesting to note that when Gerbert (Pope Sylvester II) returned from Spain around the year 1000, he introduced *apices* or counters marked with the *gubar* numerals. These counters were used to perform calculation on a board with marked columns, and a blank space stood for zero [Menninger 1969, p. 324; Smith & Karpinski 1911, pp. 117–118]. This system of reckoning is very similar to the rod numeral system.

Inscriptions of the 7th century showing the zero either in the form of a dot or a circle have been discovered in Cambodia and other parts of southeast Asia. This is about two hundred and fifty years earlier than the first epigraphic evidence in India. In discussing this significant fact, Needham [1959, pp. 10–12] posed the question of a possible stimulus from the empty blanks left for zeros on the Chinese counting boards. Such an inspiration could easily have spurred the development of a written equivalent of the rod numeral system in other countries then in contact with China.

9.9 Conclusion

We shall now recapitulate the argument developed in this chapter.

The idea that the Hindu-Arabic numeral system originated from the rod numeral system might seem strange and improbable at first. This would probably be because one is a written system, while the other computes through the use of rods.

While this difference in medium would render the rod system an unlikely precursor of the forms of the Hindu-Arabic system, it would still be perfectly logically for the underlying concept to have been transmitted from China. This is the important question for historical enquiry, for it is the discovery of the fundamental concept, rather than the forms of expression, that has lasting implications for mathematics and human civilization.

The Hindu-Arabic numeral system uses a place value notation with ten as base, and the nine signs needed are those which represent the first nine numerals. The rod numeral system which was used in China for computation since ancient times till the 17th century had an identical concept. In fact, it is the only known numeral system which is conceptually identical to the Hindu-Arabic numeral system. If the Hindu system was not an original invention, it follows that it must be transmitted from the rod system.

Since the primary use of both the Hindu-Arabic numeral system and the rod numeral system was in computation, how each system was used to add, subtract, multiply and divide, would be a good place to search for clues as to the origins of these systems. A study of their early methods showed that these were remarkably similar, despite the fact, that one system used rods and the other was a written one. This, together with the presence of other similarities in the conventions of the two numeral systems, shows that an independent origin for the Hindu-Arabic system, would be highly improbable.

It follows that the Hindu-Arabic numeral system could only have originated from the rod numeral system, which was developed centuries earlier. Transmission from China to India would not have been difficult, given the extensive use of rods in ancient China, and the considerable interaction between the Chinese and Indian civilizations, which was fostered by trade and other contacts.

Like printing, gunpowder and the magnet, three inventions which Francis Bacon said had "changed the whole face and state of things throughout the world"[4], the concept of our numeral system should rank as one of China's most significant contributions to human science and civilization.

---

[4] *Novum Organum*, book 1, aphorism 129.

PART TWO

Translation of *Sun Zi Suanjing* (The Mathematical Classic of Sun Zi)

孫

子

算

經

Preface

Master Sun says: Mathematics [governs][1] the length and breadth of the heavens and the earth; [affects] the lives of all creatures; [forms] the alpha and omega of the five constant virtues, [i.e., benevolence, righteousness, propriety, knowledge and sincerity]; [acts as] the parents for *yin* 陰 and *yang* 陽; establishes the symbols for the stars and the constellations; [manifests] the dimensions of the three luminous bodies, [i.e., sun, moon and stars]; maintains the balance of the five phases, [i.e., metal, wood, water, fire and earth]; [regulates] the beginning and the end of the four seasons; [formulates] the origins of myriad things; and [determines] the principles of the six arts, [i.e., propriety, music, archery, charioteership, calligraphy and mathematics].

[The function of mathematics] is to investigate the assembling and dispersing of the various orders [in nature], to examine the rise and fall of the two *qi* 氣,[2] [i.e., *yin* 陰 and *yang* 陽], to compute the alternating

---

[1] Square brackets are used to indicate editorial additions by way of explanation, amplification, or adaptation to the grammar of the English language.

[2] The term *qi* 氣 is somewhat analogous to the *pneuma* of the ancient Greeks and the *prana* of the ancient Hindus. It can exist in two different states, namely, *yin* and *yang*, which are generally considered as fundamental forces of the universe. The *yin* and *yang* are both opposite and complementary to each other. They interact with and dominate over each other successively in a wavelike action.

movements of the seasons, to pace out the distances [of the celestial bodies], to observe the intricate signs of the way of the heavens, to perceive the physical features of the earth, to locate the positions of the celestial and terrestrial spirits, to verify the [causes] of success and failure, to exhaust the principles of morality, and to study the temperament of life. [The field of mathematics covers] the use of the compass and the carpenter's square to regulate squares and circles, the fixing of standard measures to estimate lengths, and the establishment of measures to determine weights. [These measures] are split [to the accuracies of] *hao* 豪 and *li* 釐 [for lengths], and *shu* 黍 and *lei* 絫 [for weights].[3]

[Mathematics] has prevailed for thousands of years and has been used extensively without limitations. If one neglects its study, one will not be able to achieve excellence and thoroughness. There is indeed a great deal to master when one views mathematics in perspective. When one becomes interested in mathematics, one will be fully enriched; on the other hand, when one keeps away from [the subject], one finds oneself lacking intellectually. When one studies [mathematics] readily like a youth with an open mind, one is instantly enlightened. However if one approaches [mathematics] like an old man with an obstinate attitude, one will not be skilful in it. Therefore if one wants to learn mathematics [fruitfully], one must discipline oneself and aim for perfect concentration; it is through this way that success in learning is assured.

---

[3] For these measures, see p. 191.

Chapter 1

The measures of length begin with *hu* 忽. If one wishes to know the measure of 1 *hu*, it is [the thickness] of a strand of silk vomited by a silkworm.

10 *hu* 忽 equal 1 *si* 絲,
10 *si* equal 1 *hao* 毫,
10 *hao* equal 1 *li* 釐,
10 *li* equal 1 *fen* 分,
10 *fen* equal 1 *cun* 寸,
10 *cun* equal 1 *chi* 尺,
10 *chi* equal 1 *zhang* 丈,
10 *zhang* equal 1 *yin* 引.

50 *chi* 尺 equal 1 *duan* 端;
40 *chi* equal 1 *pi* 疋.

6 *chi* equal 1 *bu* 步;
240 *bu* equal 1 *mu* 畝;
300 *bu* equal 1 *li* 里.

The measures of weight begin with *shu* 黍.

10 *shu* equal 1 *lei* 絫,
10 *lei* equal 1 *zhu* 銖,
24 *zhu* equal 1 *liang* 兩,
16 *liang* equal 1 *jin* 斤,

30 *jin* equal 1 *jun* 鈞,

4 *jun* equal 1 *dan* 石.

The measures of capacity begin with *su* 粟.

6 *su* equal 1 *gui* 圭,

10 *gui* equal 1 *cuo* 撮,

10 *cuo* equal 1 *chao* 抄,

10 *chao* equal 1 *shao* 勺,

10 *shao* equal 1 *ge* 合,

10 *ge* equal 1 *sheng* 升,

10 *sheng* equal 1 *dou* 斗,

10 *dou* equal 1 *hu* 斛.

1 *hu* 斛 has 60,000,000 (*liu qian wan* 六千萬) *su* 粟; therefore one needs to know that

6 (*liu* 六) *su* 粟 equal 1 *gui* 圭,

10 *gui* or 60 (*liu shi* 六十) *su* equal 1 *cuo* 撮,

10 *cuo* or 600 (*liu bai* 六百) *su* equal 1 *chao* 抄,

10 *chao* or 6,000 (*liu qian* 六千) *su* equal 1 *shao* 勺,

10 *shao* or 60,000 (*liu wan* 六萬) *su* equal 1 *ge* 合,

10 *ge* or 600,000 (*liu shi wan* 六十萬) *su* equal 1 *sheng* 升,

10 *sheng* or 6,000,000 (*liu bai wan* 六百萬) *su* equal 1 *dou* 斗,

10 *dou* or 60,000,000 (*liu qian wan* 六千萬) *su* equal 1 *hu* 斛.

10 (*shi* 十) *hu* [equal] 600,000,000 (*liu yi* 六億) *su*,

100 (*bai* 百) *hu* [equal] 6,000,000,000 (*liu zhao* 六兆) *su*,

1,000 (*qian* 千) *hu* [equal] 60,000,000,000 (*liu jing* 六京) *su*,

10,000 (*wan* 萬) *hu* [equal] 600,000,000,000 (*liu gai* 六陔) *su*,

100,000 (*shi wan* 十萬) *hu* [equal] 6,000,000,000,000 (*liu zi* 六秭) *su*,

1,000,000 (*bai wan* 百萬) *hu* [equal] 60,000,000,000,000 (*liu rang* 六壤) *su*,

10,000,000 (*qian wan* 千萬) *hu* [equal] 600,000,000,000,000 (*liu gou* 六溝) *su*,

100,000,000 (*wan wan* 萬萬) *hu* is 1 *yi* 億 [which equals] 6,000,000,000,000,000 (*liu jian* 六澗) *su*,

1,000,000,000 (*shi yi* 十億) *hu* [equal] 60,000,000,000,000,000 (*liu zheng* 六正) *su*,

10,000,000,000 (*bai yi* 百億) *hu* [equal] 600,000,000,000,000,000 (*liu zai* 六載) *su*.

In the common model of large numbers (*fan da shu zhi fa* 凡大數之法):

10^8 *wan wan*[1] 萬萬 is called *yi* 億,
10^{16} *wan wan yi* 萬萬億 is called *zhao* 兆,
10^{24} *wan wan zhao* 萬萬兆 is called *jing* 京,
10^{32} *wan wan jing* 萬萬京 is called *gai* 陔,
10^{40} *wan wan gai* 萬萬陔 is called *zi* 秭,
10^{48} *wan wan zi* 萬萬秭 is called *rang* 壤,
10^{56} *wan wan rang* 萬萬壤 is called *gou* 溝,
10^{64} *wan wan gou* 萬萬溝 is called *jian* 澗,
10^{72} *wan wan jian* 萬萬澗 is called *zheng* 正,
10^{80} *wan wan zheng* 萬萬正 is called *zai* 載.

[The ratio of] the circumference of a circle to its diameter is 3 to 1.
[The ratio of] one side of a square to a diagonal is 5 to 7.
Given a diagonal, to find a side: multiply by 5 and divide by 7.
Given a side, to find a diagonal: multiply by 7 and divide by 5.

One *cun* 寸 cube of gold weighs 1 *jin* 斤,
One *cun* cube of silver weighs 14 *liang* 兩,
One *cun* cube of jade weighs 12 *liang* ,
One *cun* cube of copper weighs $7\frac{1}{2}$ *liang* ,
One *cun* cube of lead weighs $9\frac{1}{2}$ *liang*,
One *cun* cube of iron weighs 6 *liang*,
One *cun* cube of stone weighs 3 *liang*.

In the common method of computation [with rods] (*fan suan zhi fa* 凡算之法), one must first know the positions (*wei* 位) [of the rod numerals]. The units are vertical and the tens horizontal, the hundreds stand and the thousands prostrate; thousands and tens look alike and so do ten thousands and hundreds.

In the common method of multiplication (*fan cheng zhi fa* 凡乘之法), set up two positions, the upper and lower positions facing each other. If there are tens in the upper position then the correspondence is with the

---

[1] For *wan wan* and the other numbers which follow, we have translated them into powers of ten for the sake of brevity.

tens, [i.e., the units of the lower numeral are below the tens of the upper numeral]. If there are hundreds in the upper position then the correspondence is with the hundreds, [i.e., the units of the lower numeral are below the hundreds of the upper numeral]. If there are thousands in the upper position then the correspondence is with the thousands, [i.e., the units of the lower numeral are below the thousands of the upper numeral].

The upper commands the lower, [i.e., the digit of the upper numeral which is above the units of the lower numeral is multiplied by each digit of the lower numeral,] and the result is displayed in the middle position. When [multiplication with a digit of the lower numeral results in a product that] calls out tens, pass the tens over [to the left]; the remainder, [i.e., the units,] stays put [in the same column as the digit of the lower numeral].

Next remove [that digit of] the upper position which has been multiplied, and shift one place [to the right] (*tui zhi* 退之) the numeral of the lower position which has been the multiplier.[2,3] [In this manner] the numerals of the upper and lower positions are mutually multiplied till the process is completed.

In the common method of division (*fan chu zhi fa* 凡除之法), this is the reverse of multiplication. The dividend (*cheng de* 乘得 lit. product) occupies the middle position and the quotient (*chu de* 除得) is placed above it. Suppose 6 is the divisor (*fa* 法) and 100 is the dividend (*shi* 實). When 6 divides 100, it advances (*jin* 進) two places [to the left] so that it is directly below the hundreds. This implies the division of 1 by 6. In this case, the divisor (*fa*) is greater than the dividend (*shi*), so division is not possible. Therefore shift (*tui* 退) [6 to the right] so that it is below the tens. Using the divisor (*fa*) to remove the dividend (*shi*), one six [is 6] and 100 is reduced to 40, thus showing that division is possible. If the divisor (*fa*) is less than [that part of] the dividend (*shi*) [above it], it should then stay below the hundreds and should not be shifted.

---

[2] The text does not make a distinction between "multiplier" and the "multiplicand"; it uses the same phrase: *cheng qi zhe* 乘訖者.

[3] The text has the following insertion here: "6 is not an accumulation [of rods] and 5 is not a single [rod]". We have omitted this in the translation, as this sentence is misplaced here. It should have been inserted at the end of the paragraph about rod numerals (p. 193), and this would have agreed with the context of a similar sentence in *Xiahou Yang suanjing* (see [Qian ed. 1963, p. 558]). (See p. 47)

It follows that if the units of the divisor (*fa*) are below the tens [of the dividend], the place value [of the digit of the quotient] is tens, if they are below the hundreds, the place value [of the digit of the quotient] is hundreds.[4] The rest of the method is the same as multiplication.

As for the remainder of the dividend (*shi*), this is assigned to the divisor (*yi fa ming zhi* 以法命之) such that the divisor (*fa*) is called the denominator (*mu* 母) and the remaining dividend (*shi*) the numerator (*zi* 子).

> To find coarse rice (*li mi* 糲米) from millet (*su* 粟): multiply by 3 and divide by 5.
> To find millet from coarse rice: multiply by 5 and divide by 3.
> To find cooked [coarse] rice (*fan* 飯) from coarse rice: multiply by 5 and divide by 2.
> To find cooked coarse rice from millet: multiply by 6 and divide by 4.
> To find coarse rice from cooked coarse rice: multiply by 2 and divide by 5.
> To find cooked [refined] rice from refined rice (*zuo mi* 鑿米): multiply by 8 and divide by 4.

> For 1 part out of [5] 10: multiply by 2 and divide by 20.
> For 2 parts out of 10: multiply by 4 and divide by 20.
> For 3 parts out of 10: multiply by 6 and divide by 20.
> For 4 parts out of 10: multiply by 8 and divide by 20.
> For 5 parts out of 10: multiply by 10 and divide by 20.
> For 6 parts out of 10: multiply by 12 and divide by 20.
> For 7 parts out of 10: multiply by 14 and divide by 20.
> For 8 parts out of 10: multiply by 16 and divide by 20.
> For 9 parts out of 10: multiply by 18 and divide by 20.

> For 1 part out of 9: multiply by 2 and divide by 18.
> For 1 part out of 8: multiply by 2 and divide by 16.
> For 1 part out of 7: multiply by 2 and divide by 14.

---

[4] To elucidate the concept of blank spaces, Li Chunfeng 李淳風 inserted the following: "If [the numeral of] the upper position has an empty space, this means that the divisor has been shifted two places [to the right]". This is the only comment of Li Chunfeng in the book. See pp. 30 & 65.

[5] *Jian* 減 is translated here as "out of" instead of the usual meaning "subtracted from".

For 1 part out of 6: multiply by 2 and divide by 12.
For 1 part out of 5: multiply by 2 and divide by 10.

[1]⁶ Nine nines are 81, find the amount when this is multiplied by itself.
Answer:⁷ 6,561.
Method:⁸ Set up the two positions: [upper and lower]. The upper 8 calls the lower 8: eight eights are 64, so put down 6,400 in the middle position. The upper 8 calls the lower 1: one eight is 8, so put down 80 in the middle position. Shift the lower numeral one place [to the right] and put away the 80 in the upper position. The upper 1 calls the lower 8: one eight is 8, so put down 80 in the middle position. The upper 1 calls the lower 1: one one is 1, so put down 1 in the middle position. Remove the numerals in the upper and lower positions leaving 6,561 in the middle position.

[2]   If 6,561 is divided among 9 persons, find how much each gets.
Answer: 729.
Method: First set 6,561 in the middle position to be the *shi* 實 (dividend). Below it, set 9 persons to be the *fa* 法 (divisor). Put down 700 in the upper position. The upper 7 calls the lower 9: seven nines are 63, so remove 6,300 from the numeral in the middle position. Shift the numeral in the lower position one place [to the right] and put down 20 in the upper position. The upper 2 calls the lower 9: two nines are 18, so remove 180 from the numeral in the middle position. Once again shift the numeral in the lower position one place [to the right], and put down 9 in the upper position. The upper 9 calls the lower 9: nine nines are 81, so remove 81 from the numeral in the middle position. There is now no numeral in the middle position. Put away the numeral in the lower position. The result in the upper position is what each person gets.

Similar procedures can be applied from eight eights are 64 to one one is 1.

---

⁶ There are only two problems in this chapter and we have numbered them 1 and 2. For the problems of Chapters 2 & 3, we have followed Qian Baocong' s numbering.
⁷ The literal translation is "answer says" (*da yue* 答曰).
⁸ The literal translation is "method says" (*shu yue* 術曰).

Eight nines are 72; multiply this by itself to obtain 5,184. When this is divided among 8 persons, each person gets 648.

Seven nines are 63; multiply this by itself to obtain 3,969. When this is divided among 7 persons, each person gets 567.

Six nines are 54; multiply this by itself to obtain 2,916. When this is divided among 6 persons, each person gets 486.

Five nines are 45; multiply this by itself to obtain 2,025. When this is divided among 5 persons, each person gets 405.

Four nines are 36; multiply this by itself to obtain 1,296. When this is divided among 4 persons, each person gets 324.

Three nines are 27; multiply this by itself to obtain 729. When this is divided among 3 persons, each person gets 243.

Two nines are 18; multiply this by itself to obtain 324. When this is divided among 2 persons, each person gets 162.

One nine is 9; multiply this by itself to obtain 81. One person gets 81.

[The sum of the products from] nine nines listed above[9] is 405;[10] multiply this by itself to obtain 164,025. When this is divided among 9 persons, each person gets 18,225.

Eight eights are 64; multiply this by itself to obtain 4,096. When this is divided among 8 persons, each person gets 512.

Seven eights are 56; multiply this by itself to obtain 3,136. When this is divided among 7 persons, each person gets 448.

Six eights are 48; multiply this by itself to obtain 2,304. When this is divided among 6 persons, each person gets 348.

Five eights are 40; multiply this by itself to obtain 1,600. When this is divided among 5 persons, each person gets 320.

Four eights are 32; multiply this by itself to obtain 1,024. When this is divided among 4 persons, each person gets 256.

Three eights are 24; multiply this by itself to obtain 576. When this is divided among 3 persons, each person gets 192.

---

[9] We have used the word "above" for obvious reasons. The literal translation is "on the right", since a Chinese book is read linearly from top to bottom and in the direction from right to left.

[10] I.e., $81 + 72 + 63 + 54 + 45 + 36 + 27 + 18 + 9 = 405$.

Two eights are 16; multiply this by itself to obtain 256. When this is divided among 2 persons, each person gets 128.

One eight is 8; multiply this by itself to obtain 64. One person gets 64.

[The sum of the products from] eight eights listed above is 288; multiply this by itself to obtain 82,944. When this is divided among 8 persons, each person gets 10,368.

Seven sevens are 49; multiply this by itself to obtain 2,401. When this is divided among 7 persons, each person gets 343.

Six sevens are 42; multiply this by itself to obtain 1,764. When this is divided among 6 persons, each person gets 294.

Five sevens are 35; multiply this by itself to obtain 1,225. When this is divided among 5 persons, each person gets 245.

Four sevens are 28; multiply this by itself to obtain 784. When this is divided among 4 persons, each person gets 196.

Three sevens are 21; multiply this by itself to obtain 441. When this is divided among 3 persons, each person gets 147.

Two sevens are 14; multiply this by itself to obtain 196. When this is divided among 2 persons, each person gets 98.

One seven is 7; multiply this by itself to obtain 49. One person gets 49.

[The sum of the products from] seven sevens listed above is 196; multiply this by itself to obtain 38,416. When this is divided among 7 persons, each person gets 5,488.

Six sixes are 36; multiply this by itself to obtain 1,296. When this is divided among 6 persons, each person gets 216.

Five sixes are 30; multiply this by itself to obtain 900. When this is divided among 5 persons, each person gets 180.

Four sixes are 24; multiply this by itself to obtain 576. When this is divided among 4 persons, each person gets 144.

Three sixes are 18; multiply this by itself to obtain 324. When this is divided among 3 persons, each person gets 108.

Two sixes are 12; multiply this by itself to obtain 144. When this is divided among 2 persons, each person gets 72.

One six is 6; multiply this by itself to obtain 36. One person gets 36.

[The sum of the products from] six sixes listed above is 126; multiply this by itself to obtain 15,876. When this is divided among 6 persons, each person gets 2,646.

Five fives are 25; multiply this by itself to obtain 625. When this is divided among 5 persons, each person gets 125.

Four fives are 20; multiply this by itself to obtain 400. When this is divided among 4 persons, each person gets 100.

Three fives are 15; multiply this by itself to obtain 225. When this is divided among 3 persons, each person gets 75.

Two fives are 10; multiply this by itself to obtain 100. When this is divided among 2 persons, each person gets 50.

One five is 5; multiply this by itself to obtain 25. One person gets 25.

[The sum of the products from] five fives listed above is 75; multiply this by itself to obtain 5,625. When this is divided among 5 persons, each person gets 1,125.

Four fours are 16; multiply this by itself to obtain 256. When this is divided among 4 persons, each person gets 64.

Three fours are 12; multiply this by itself to obtain 144. When this is divided among 3 persons, each person gets 48.

Two fours are 8; multiply this by itself to obtain 64. When this is divided among 2 persons, each person gets 32.

One four is 4; multiply this by itself to obtain 16. One person gets 16.

[The sum of the products from] four fours listed above is 40; multiply this by itself to obtain 1,600. When this is divided among 4 persons, each person gets 400.

Three threes are 9; multiply this by itself to obtain 81. When this is divided among 3 persons, each person gets 27.

Two threes are 6; multiply this by itself to obtain 36. When this is divided among 2 persons, each person gets 18.

One three is 3; multiply this by itself to obtain 9. One person gets 9.

[The sum of the products from] three threes listed above is 18; multiply this by itself to obtain 324. When this is divided among 3 persons, each person gets 108.

Two twos are 4; multiply this by itself to obtain 16. When this is divided among 2 persons, each person gets 8.

Two one is 2; multiply this by itself to obtain 4. One person gets 4.

[The sum of the products from] two twos listed above is 6; multiply this by itself to obtain 36. When this is divided among 2 persons, each person gets 18.

One one is 1; multiply this by itself to obtain 1. Multiplication by 1 does not create an increment.

The sum of all the above from nine nines to one one is 1,155;[11] multiply this by itself to obtain 1,334,025. When this is divided among 9 persons, each person gets 148,225.

Multiply 12 by 9 to obtain 108. When this is divided among 6 persons, each person gets 18.

Multiply 36 by 27 to obtain 972. When this is divided among 18 persons, each person gets 54.

Multiply 108 by 81 to obtain 8,748. When this is divided among 54 persons, each person gets 162.

Multiply 324 by 243 to obtain 78,732. When this is divided among 162 persons, each person gets 486.

Multiply 972 by 729 to obtain 708,588. When this is divided among 486 persons, each person gets 1,458.

Multiply 2,916 by 2,187 to obtain 6,377,292. When this is divided among 1,458 persons, each person gets 4,374.

Multiply 8,748 by 6,561 to obtain 57,395,628. When this is divided among 4,374 persons, each person gets 13,122.

Multiply 26,244 by 19,683 to obtain 516,560,652. When this is divided among 13,122 persons, each person gets 39,366.

Multiply 78,732 by 59,049 to obtain 4,649,045,868. When this is divided among 39,366 persons, each person gets 118,098.

Multiply 236,196 by 177,147 to obtain 41,841,412,812. When this is divided among 118,098 persons, each person gets 354,294.

Multiply 708,588 by 531,441 to obtain 376,572,715,308. When this is divided among 354,294 persons, each person gets 1,062,882.

---

[11] I.e., 405 + 288 + 196 + 126 + 75 + 40 +18 + 6 + 1 = 1,155.

Chapter 2

[1] Now there is a fraction[1] $\frac{12}{18}$; reduce (*yue* 約) it to find its [simplest] form.

Answer: $\frac{2}{3}$.

Method: Put down 18 in the lower position and 12 in the upper position. Set the numerals in two other positions for the purpose of subtracting the smaller from the larger to derive 6 as the *deng shu* 等數 (greatest common divisor lit. equal number). Use it as the divisor (*fa* 法) for reducing (*yue zhi* 約之) [the fraction] to obtain the answer.

[2] Now there are fractions $\frac{1}{3}$ and $\frac{2}{5}$, find their sum (*he* 合).

Answer: $\frac{11}{15}$.

Method: Put down denominators (*fen* 分) 3 and 5 on the right and numerators (*zhi* 之) 1 and 2 on the left. Multiply a numerator and the other denominator (*mu hu cheng zhi* 母互乘之) to obtain 6 for the fraction $\frac{2}{5}$ and 5 for the fraction $\frac{1}{3}$. Add to obtain 11, which becomes the dividend (*shi* 實). Multiply the two denominators on the right to

---

[1] We have taken the liberty to translate fractions into our own notational form. In the text, the fraction, $\frac{12}{18}$, is written as *yi shi ba fen zhi yi shi er* 一十八分之一十二 which means "12 parts out of 18". The reader is reminded that in the computation of fractions through rod numerals, they were expressed in a similar notation as ours. See p. 79.

give 15, which becomes the divisor (*fa* 法)[2]. [The dividend] is less than the divisor, so assign it to the divisor (*yi fa ming zhi* 以法命之) to get the answer.

[3] Now there is a fraction $\frac{8}{9}$ from which $\frac{1}{5}$ is subtracted (*jian* 减). Find the remainder.

Answer: $\frac{31}{45}$.

Method: Put down denominators 9 and 5 on the right and numerators 8 and 1 on the left. Multiply a numerator and the other denominator to obtain 9 for the fraction $\frac{1}{5}$ and 40 for the fraction $\frac{8}{9}$. Subtract the smaller from the larger to obtain a remainder, 31, which becomes the dividend (*shi*). Multiply the denominators to give 45, which becomes the divisor (*fa*). [The dividend] is less than the divisor, so assign it to the divisor (*yi fa ming zhi* 以法命之) to get the answer.

[4] Now there are fractions $\frac{1}{3}$, $\frac{2}{3}$ and $\frac{3}{4}$. Find the amounts to be subtracted from the larger [fractions] and that to be added to the smallest [in order to obtain] the average value (*ping* 平), and find this value.

Answer: Subtract 2 [twelfths] from $\frac{3}{4}$ and 1 [twelfth] from $\frac{2}{3}$; the sum [of the subtrahends] is added to $\frac{1}{3}$. Each gives the average value of $\frac{7}{12}$.

Method: Put down denominators 3, 3 and 4 on the right and numerators 1, 2 and 3 on the left. Multiply a numerator and the other denominators; add [the products] to obtain 63 and put this on the right calling it *ping shi* 平實 (lit. average dividend). Multiply the denominators to obtain 36, which becomes the divisor (*fa*). Use 3, which is the number of the displayed [fractions], to multiply [each of the products] before their addition and the divisor (*fa*). 9 is derived as the *deng shu* 等數 (greatest common divisor) and this reduces [the fractions]. Subtract 2 [twelfths] from $\frac{3}{4}$ and 1 [twelfth] from $\frac{2}{3}$. The sum [of the subtrahends] is added to $\frac{1}{3}$. Each gives the average value of $\frac{7}{12}$.

---

[2] For an account of the technical words *shi* and *fa*, see pp. 64–65.

[5]   Now there is 1 *dou*[3] 斗 of millet (*su* 粟), find the equivalent amount of coarse rice (*li mi* 糲米).

Answer: 6 *sheng* 升.

Method: Put down 1 *dou* or 10 *sheng* of millet. Multiply this by the proportional value of coarse rice, 30, to obtain 300 *sheng* which becomes the dividend (*shi*). The proportional value of millet, 50, becomes the divisor (*fa*). The answer is obtained on division.

[6]   Now there are 2 *dou* 1 *sheng* of millet, find the equivalent amount of polished rice (*bai mi* 粺米).

Answer: 1 *dou* $1\frac{17}{50}$ *sheng*.

Method: Put down 21 *sheng* of millet. Multiply this by the proportional value of polished rice, 27, to obtain 567 *sheng* which becomes the dividend (*shi*). The proportional value of millet, 50, becomes the divisor (*fa*). Perform the division. The remainder is assigned to the divisor to form a fraction.

[7]   Now there are 4 *dou* 5 *sheng* of millet, find the equivalent amount of refined rice (*zuo mi* 繫米).

Answer: 2 *dou* $1\frac{3}{5}$ *sheng*.

Method: Put down 45 *sheng* of millet. Divide the proportional value of refined rice, 24, by 2 to give 12 and multiply [the amount of millet] by this to obtain 540 *sheng* which becomes the dividend (*shi*). Divide the proportional value of millet, 50, by 2 to give 25 which becomes the divisor (*fa*). Perform the division. The remainder [and the divisor] are divided by the *deng shu* 等數 (greatest common divisor) resulting in a fraction [of the simplest form].

[8]   Now there are 7 *dou* 9 *sheng* of millet, find the equivalent amount of imperial rice (*yu mi* 御米).

Answer: 3 *dou* 3 *sheng* 1 *ge* 合 8 *shao* 勺.

Method: Put down 7 *dou* 9 *sheng* and multiply this by the proportional value of imperial rice, 21, to obtain 1,659 *sheng* which becomes the dividend (*shi*). Divide by the proportional value of millet, 50, to get the answer.

---

[3] Refer to Ch. 1 & Sect. 6 for tables of measures.

[9]   Now there is the base of a house, 3 *zhang* 丈 in the north-south direction and 6 *zhang* in the east-west direction, which is to be laid with bricks. Find the number [of bricks] required if 5 pieces cover an area of 2 *chi* 尺.

Answer: 4,500 pieces.

Method: Put down the east-west [length] of 6 *zhang* and multiply it by the north-south [length] of 3 *zhang* to obtain 1,800 *chi*. Multiply by 5 to give 9,000 *chi*, and divide by 2 to get the answer.

[10] Now there is a cylindrical [*yuan* 圓 lit. circular] cellar whose base circumference is 286 *chi* and whose depth is 3 *zhang* 6 *chi*. How much millet can it hold?

Answer: 151,474 *hu* 斛 $7\frac{11}{27}$ *sheng*.

Method: Put down the circumference,[4] 286 *chi*, multiply this by itself to give 81,796 *chi*. Multiply by the depth, 3 *zhang* 6 *chi*, to obtain 2,944,656 and divide by 12 to obtain 245,388 *chi*. Divide by 1 *chi* 6 *cun* 寸 2 *fen* 分, which is the measure of one *hu*, to get the answer.

[11] Now there is a rectangular cellar 4 *zhang* 6 *chi* wide, 5 *zhang* 4 *chi* long and 3 *zhang* 5 *chi* deep. How much millet can it hold?

Answer: 53,666 *hu* 6 *dou* $6\frac{2}{3}$ *sheng*.[5]

Method: Put down width, 4 *zhang* 6 *chi*, and length, 5 *zhang* 4 *chi*; multiply them to obtain 2,484 *chi*. Multiply by the depth 3 *zhang* 5 *chi*, to obtain 86,940 *chi*. Divide by 1 *chi* 6 *cun* 2 *fen*, which is the measure of one *hu*, to get the answer.

[12] Now there is a cylindrical cellar whose circumference is 5 *zhang* 4 *chi* and whose depth is 1 *zhang* 8 *chi*. How much millet can it hold?

Answer: 2,700 *hu*.

Method: First put down the circumference, 5 *zhang* 4 *chi*, and multiply

---

[4] In the translation we have followed the technical style of the text and have not written the sentence as "put down [the length of] the circumference". This mode of translation is used frequently throughout the book.

[5] Qian [ed. 1963 p. 299, n. 1] says, "In '$\frac{2}{3}$ *sheng*', all editions mistook the [numerator] '2' for '1'. It is now corrected."

this by itself to give 2,916 *chi*. Multiply by the depth, 1 *zhang* 8 *chi*, to obtain 52,488 *chi* and divide by 12 to obtain 4,374 *chi*. Divide by 1 *chi* 6 *cun* 2 *fen*, which is the measure of one *hu*, to get the answer.

[13] Now there is a circular field whose circumference is 300 *bu* 步 and whose diameter is 100 *bu*. Find the area.

Answer: 31 *mu* 畝 and an odd lot (*ji* 奇) of 60 *bu*.

Method: First put down the circumference, 300 *bu*, halve it to obtain 150 *bu*. Next put down the diameter, 100 *bu*, halve it to obtain 50 *bu*. Multiply them to obtain 7,500 *bu* and divide by 240 *bu*, which is the measure of one *mu*, to get the answer.

Another method: Multiply the circumference by itself to obtain 90,000 *bu* and divide by 12 to give 7,500 *bu*. Divide by the measure of one *mu* to get the amount in *mu*.

Another method: Multiply the diameter by itself to obtain 10,000 and multiply by 3 to give 30,000 *bu*. Divide by 4 to obtain 7,500 *bu*. Divide by the measure of one *mu* to get the amount in *mu*.

[14] Now there is a square field with a mulberry tree at the centre. The distance of one corner [of the field] to the mulberry tree is 147 *bu*. Find the area of the field.

Answer: 1 *qing*[6] 頃 83 *mu* and an odd lot of 180 *bu*.

Method: Put down the distance from the corner to the mulberry tree, 147 *bu*, and double it to obtain 294 *bu*. Multiply by 5 to give 1,470 *bu* and divide by 7 to obtain 210 *bu*. Multiply this by itself to obtain 44,100 *bu* and divide by 240 *bu* to get the answer.

[15] Now there is a cube of wood of side 3 *chi*. If a cube of side 5 *cun* makes one pillow, find the number [of pillows].

Answer: 216 pieces.

Method: Put down a side of the cube, 3 *chi*, multiply by itself to obtain 9 *chi*. Multiply by the height, 3 *chi*, to obtain 27 *chi*. Multiply this by [the rate of] 8 pillows per *chi* [cube] of wood to get the answer.

---

[6] 1 *qing* = 100 *mu*.

[16] Now there is a rope 5,794 *bu* long which is used to make a square. Find a side [of the square].

Answer: 1,448 *bu* 3 *chi*.

Method: Put down the length of the rope, 5,794 *bu*, and divide by 4 to give 1,448 *bu* and a remainder of 2 *bu*. Multiply [the latter] by 6 to obtain 1 *zhang* 2 *chi* and divide by 4 to obtain 3 *chi*. The results taken together yield the answer.

[17] Now there is an embankment whose lower width is 5 *zhang*, upper width 3 *zhang*, height 2 *zhang*, and length 60 *chi*. If one cube has [a volume] of 1,000 *chi*, find the number [of cubes].

Answer: 48 cubes.

Method: Put down the upper width of the embankment, 3 *zhang*, and its lower width, 5 *zhang*. Add them to obtain 8 *zhang* and halve this to give 4 *zhang*. Multiply this by the height, 2 *zhang*, to obtain 800 *chi*. Multiply by the length, 60 *chi*, to obtain 48,000 and divide by 1,000 *chi* to get the answer.

[18] Now there is a drain 10 *zhang* wide, 5 *zhang* deep and 20 *zhang* long. If one cube has [a volume of] 1,000 *chi*, find the number [of cubes].

Answer: 1,000 cubes.

Method: Put down the width, 10 *zhang*, and multiply by the depth, 5 *zhang*, to obtain 5,000 *chi*. Next multiply by the length, 20 *zhang*, to give 1,000,000 *chi*. Divide by 1,000 to get the answer.

[19] Now there is an area of 234,567 *bu*, find [one side of] the square.

Answer: $484\frac{311}{968}$ *bu*.

Method: Put down the area, 234,567 *bu*, as *shi* 實 (lit. dividend). Next use one rod as *xia fa* 下法 (lit. lower divisor). [Move this rod from below] the units in *bu* [of the *shi*] overpassing one place (*chao yi wei* 超一位) to reach the hundreds[7] and stop (*zhi bai er zhi* 至百而止). Put down 400 as *shang* 商 (lit. quotient) above the *shi*. Next put down 40,000 below the *shi* and above the *xia fa* and call it *fang fa* 方法

---

[7] The word "hundreds" should be replaced by "ten thousands". See p. 95.

(square divisor). Assign (*ming* 命) the *shang*, 400, in the top position to it [in order] to subtract from the *shi*. After subtraction double the *fang fa*. Shift (*tui* 退) the *fang fa* [to the right] by one place and the *xia fa* by two places. Put down 80 in the top position as *shang* next to the previous *shang*. Also put down 800 below the *fang fa* and above the *xia fa* and call it *lian fa* 廉法 (lit. side divisor). Assign (*ming*) the *shang*, 80, in the top position to each of the *fang* and *lian* [in order] to subtract from the *shi*. After subtraction, double the *lian fa* and join it to the *fang fa* above. Shift the *fang fa* [to the right] by one place and the *xia fa* by two places. Put down 4 in the top position as *shang* next to the previous [*shang*]. Also put down 4 below the *fang fa* and above the *xia fa* and call it *yu fa* 隅法 (lit. corner divisor). Assign (*ming*) the *shang*, 4, in the top position to each of the *fang*, *lian* and *yu* [in order] to subtract from the *shi*. After subtraction, double the *yu fa* and join this to the *fang fa*.[8] The *shang* in the top position has 484 and the *fa* in the lower position has 968, while the remainder is 311. Hence [one side of] the square is $484\frac{311}{968}$ *bu*.

[20] Now there is an area of 35,000 *bu*, find [the circumference of] the circle.

Answer: $648\frac{96}{1296}$ *bu*.

Method: Put down the area, 35,000 *bu*, and multiply by 12 to obtain 420,000, which is the *shi*. Next use one rod as *xia fa*. [Move this rod from below] the units in *bu* [of the *shi*] overpassing one place to reach the hundreds[9] and stop. Put down 600 as *shang* above the *shi*. Next put down 60,000 below the *shi* and above the *xia fa* and call it *fang fa*. Assign the *shang*, 600, in the top position to it [in order] to subtract from the *shi*. After subtraction double the *fang fa*. Shift the *fang fa* [to the right] by one place and the *xia fa* by two places. Put down 40 in the top position as *shang* next to the previous *shang*. Also put down

---

[8] Qian [ed. 1963, p. 303. n. 2] says, "The statement 'double the *yu fa* and join this to the *fang fa*' is missing in the Southern Song version and the [*Yongle*] *da dian* edition. It was added by Dai Zhen following the given method. The present version follows [this amendment]."

[9] See fn. 7.

400 below the *fang fa* and above the *xia fa* and call it *lian fa*. Assign the *shang*, 40,[10] in the top position to each of the *fang* and *lian* [in order] to subtract from the *shi*. After subtraction, double the *lian fa* and join it to the *fang fa*. Shift the *fang fa* [to the right] by one place and the *xia fa* by two places. Put down 8 in the top position as *shang* next to the previous *shang*. Also put down 8 below the *fang fa* and above the *xia fa* and call it *yu fa*. Assign the *shang*, 8, in the top position to each of the *fang*, *lian* and *yu* [in order] to subtract from the *shi*. After subtraction, double the *yu fa* and join this to the *fang fa*. The *shang* in the top position has 648 and the *fa* in the lower position has 1,296,[11] while the remainder is 96. Hence [the circumference of] the circle[12] is $648\frac{96}{1296}$ *bu*.

[21] Now there is a spherical segment (*qiu tian* 丘田) whose [base] circumference (*zhou* 周) is 639 *bu*, and the diameter (*jing* 徑) [of the sphere] is 380 *bu*. Find the [area of the] spherical segment.

Answer: 2 *qing* 52 *mu* 225 *bu*.

Method: Halve the circumference to obtain 319 *bu* 5 *fen*, halve the diameter to obtain 190 *bu*. Multiply [the numbers in] the two positions to give 60,705 *bu*. Divide by the measure of one *mu* to get the answer.[13]

[22] Now there is a city wall of upper width 2 *zhang*, lower width 5 *zhang* 4 *chi*, height 3 *zhang* 8 *chi* and length 5,550 *chi* to be constructed. The work capacity of a person in autumn is 300 *chi*. Find the manpower needed.

---

[10] Qian [ed. 1963, p. 303. n. 4] says, "The Southern Sung version does not have the figure ' 40'. It is now added according to the [*Tianlu linlang congshu*] palace edition."

[11] Qian [ed. 1963, p. 303 n. 6] says, "The Southern Song version and [*Yongle*] *dadian* mistook the figure '1,296' for '1,297'. It is now corrected according to Dai Zhen's collation.

[12] In the text "circle" is wrongly printed as "square".

[13] The interpretation of this problem is based on the following assumptions: (a) *qiu tian* (lit. mound or hillock) means spherical segment (Needham [1959, p. 99] gives this translation), (b) "circumference" (*zhou*) is the base circumference of the spherical segment, (c) "diameter" (*jing*) is the diameter of the sphere, and (d) it is required to find the surface area of the spherical segment. On this interpretation, the answer and method of the problem are inaccurate. See also p. 128.

Answer: 26,011 manpower.

Method: Add the upper and lower widths to obtain 74 *chi*, halve this to give 37 *chi*. Multiply by the height to obtain 1,406 *chi*. Next multiply by the length to obtain a volume of 7,803,300 *chi*. Divide by 300 *chi*, which is the work capacity of a person in autumn, to get the answer.

[23] Now there is a canal of length 29 *li* 里 104 *bu*, upper width 1 *zhang* 2 *chi* 6 *cun*, lower width 8 *chi* and depth 1 *zhang* 8 *chi* to be constructed. The work capacity of a person in autumn is 300 *chi*. Find the manpower needed.

Answer: 32,645 manpower with a remainder (*bu jin* 不盡) of 69 *chi* 6 *cun*.

Method: Put down the amount in *li* and multiply by 300 *bu*; enter this into the remaining amount in *bu* and multiply by 6 to obtain 52,824 *chi*. Add the upper and lower widths to give 2 *zhang* 6 *cun*, halve this and multiply by the depth to obtain 185 *chi* 4 *cun*. Multiply by the length to obtain 9,793,569 *chi* 6 *cun*. Divide by 300 *chi*, which is the work capacity of one person, to obtain the answer.

[24] Now there are 6,930 *qian* 錢 (coins) which are to be divided among 216 persons in 9 shares [as follows]: 81 persons get 2 shares each, 72 persons get 3 shares each, and 63 persons get 4 shares each. Find how much each person gets.

Answer: [For those with] 2 shares, each person gets 22 *qian*; [for those with] 3 shares, each person gets 33 *qian*; [and for those with] 4 shares, each person gets 44 *qian*.

Method: First put down 81 persons in the upper position, 72 persons in the next position and 63 persons in the lower position. Multiply the upper numeral by 2 to obtain 162, multiply the next numeral by 3 to obtain 216, and multiply the lower numeral by 4 to obtain 252. Add the numerals in the three positions to give 630, which becomes the divisor (*fa*). Next put down 6,930 *qian* in three positions. Multiply the upper numeral by 162 to obtain 1,122,660, multiply the middle numeral by 216 to obtain 1,496,880, and multiply the lower numeral by 252 to obtain 1,746,360. Each becomes the dividend (*shi*). Divide each by the divisor (*fa*), 630, to obtain 1,782 in the upper position,

2,376 in the middle position and 2,772 in the lower position. Divide each by the [corresponding] number of persons to get the answer.

[25] Now there are five noblemen of five different ranks sharing a total of 60 tangerines. If each person has 3 tangerines more [than the next person of a lower rank], find how many tangerines each man receives. Answer: *Gong*[14] 公 (duke) gets 18, *Hou* 侯 (marquis) gets 15, *Bo* 伯 (earl) gets 12, *Zi* 子 (viscount) gets 9, and *Nan* 男 (baron) gets 6.
Method: First put down in the bottom position 3 tangerines, which is the amount to be added for each successive rank. [Put down] 6 tangerines in the next position [above], 9 tangerines in the next position, 12 tangerines in the next position, and 15 tangerines in the top position. Add them to obtain 45. Subtract this from 60 and divide the remainder by the number of persons to give 3 tangerines for each person. Add this to each [of the numerals] before they were added up (*bu bing zhe* 不并者) to obtain 18 in the top position which is *Gong'*s share, 15 in the next position [below] which is *Hou'*s share, 12 in the next position which is *Bo'*s share, 9 in the next position which is *Zi'*s share, and 6 in the bottom position which is *Nan'*s share.

[26] Now there are three persons A, B and C who hold certain sums of money. A says to B and C, "If one half of each of your money is added to mine, the result is 90." B says to A and C, "If one half of each of your money is added to mine, the result is 70." C says to A and B, "If one half of each of your money is added to mine, the result is 56." How much money does each of the three men hold originally?
Answer: A 72, B 32 and C 4.
Method: First put down in [three] positions the amounts declared by the three persons, and multiply each by 3 obtaining 270 for A, 210 for

---

[14] In the bureaucratic hierarchy of traditional China, numerous official titles were conferred on members of the imperial family and distinguished personalities. As early as the Zhou dynasty, the demarcation of the various ranks was well defined. There were twelve titles of nobility, comprising three varieties of princes (*wang* 王), five of dukes (*gong*), and one each of marquis (*hou*), earl (*bo*), viscount (*zi*) and baron (*nan*). As we can see in this problem, the ranks remained unchanged even during the time of Sun Zi. See p. 165.

B and 168 for C. Halving each yields 135 for A, 105 for B and 84 for C. Once again put down 90 for A, 70 for B and 56 for C, and halve each of them. Subtract [the latter[15]] A and B from [the previous] C, [the latter] A and C from [the previous] B, and [the latter] B and C from [the previous] A, to yield the original amounts of money held by each person.

[27] Now there is a girl who weaves skilfully. Each day she doubles the amount of weaving [done on the previous day]. In 5 days she has woven 5 *chi*. Find the amount she weaves each day.

Answer: First day $1\frac{19}{31}$ *cun*, next day $3\frac{7}{31}$ *cun*, next day $6\frac{14}{31}$ *cun*, next day 1 *chi* $2\frac{28}{31}$ *cun*, next day 2 *chi* $5\frac{25}{31}$ *cun*.

Method: Put down the [daily weaving] proportional parts (*lie cui* 列衰) and add them to obtain 31, which becomes the divisor (*fa*). Multiply [the numerals] before they were added up (*wei bing zhe* 未并者) by 5 *chi*, and let each [product] be the dividend (*shi*). Divide by the divisor (*ru fa er yi* 如法而一) to get the answer.

[28] Now there is a gang of robbers who stole an unknown quantity of thin silk from a warehouse. In the distribution of the silk among themselves, it is heard that if each person is given 6 *pi* 匹, there is a surplus (*ying* 盈) of 6 *pi*, and if each person is given 7 *pi*, there is a deficit (*bu zu* 不足) of 7 *pi*. Find the number of persons and the amount of thin silk.

Answer: 13 robbers and 84 *pi* of thin silk.

Method: First put down on the upper right, the amount each person gets, 6 *pi*, and on the lower right, the surplus, 6 *pi*. After this put down on the upper left, the amount each person gets, 7 *pi*, and on the lower left, the deficit, 7 *pi*. Cross-multiply (*wei cheng* 維乘) and add the results to obtain the quantity of thin silk. Add the surplus and deficit to obtain the number of persons.

---

[15] For an explanation of the terms "latter" and "previous", see p. 133.

Chapter 3

[1] Now there are 9 households A, B, C, D, E, F, G, H and I who collectively send their land tax [in kind]. A pays 35 *hu* 斛 [of grains], B 46 *hu*[1], C 57 *hu*, D 68 *hu*, E 79 *hu*, F 80 *hu*, G 100 *hu*, H 210 *hu* and I 325 *hu*. All 9 households send a total of 1,000 *hu* to pay their land tax. Out of this 200 *hu* is deducted for transport, find [the actual tax] each household pays.

Answer: A 28 *hu*, B 36 *hu* 8 *dou* 斗, C 45 *hu* 6 *dou*[2], D 54 *hu* 4 *dou*, E 63 *hu* 2 *dou*, F 64 *hu*, G 80 *hu*, H 168 *hu* and I 260 *hu*.

Method: Put down A' s payment of 35 *hu*, multiply by 4 to obtain 140 *hu* and divide by 5 to give 28 *hu*. B pays 46 *hu*, multiply by 4 to obtain 184 *hu* and divide by 5 to give 36 *hu* 8 *dou*. C pays 57 *hu*, multiply by 4 to obtain 228 *hu* and divide by 5 to give 45 *hu* 6 *dou*. D pays 68 *hu*, multiply by 4 to obtain 272 *hu* and divide by 5 to give 54 *hu* 4 *dou*. E pays 79 *hu*, multiply by 4 to obtain 316 *hu* and divide by 5 to give 63 *hu* 2 *dou*. F pays 80 *hu*, multiply by 4 to obtain 320 *hu* and divide by 5 to give 64 *hu*. G pays 100 *hu*, multiply by 4 to obtain 400 *hu* and

---

[1] Qian [ed. 1963, p. 309, n. 1] says, "The Southern Song version omitted the '6' in '46 *hu*' . Dai Zhen made the amendment according to the [*Yongle*] *dadian* edition."

[2] Qian [ed. 1963, p. 309, n. 2] says, "The Southern Song version omitted '6 *dou*'. Dai Zhen made the amendment according to the [*Yongle*] *dadian* edition."

divide by 5 to give 80 *hu*. H pays 210 *hu*, multiply by 4 to obtain 840 *hu* and divide by 5 to give 168 *hu*. I pays 325 *hu*, multiply by 4 to obtain 1,300 *hu* and divide by 5 to give 260 *hu*.

[2] Now there are 15,000,000 men of whom 400,000 are conscripted as soldiers. Find the rate of the number of men to one soldier.

Answer: 37 men 5 *fen* 分.

Method: Put down the number of men, 15,000,000 as the dividend (*shi*) and 400,000 soldiers as the divisor (*fa*). Divide (*shi ru fa* 實如法, lit. divide the *shi* by the *fa*) to obtain the answer.

[3] Now there is a [conical] pile of millet on a levelled ground. The circumference of the base is 3 *zhang* 丈 6 *chi* 尺 and the height is 4 *chi* 5 *cun* 寸. Find the amount of millet.

Answer: 100 *hu*.

Method: Put down the circumference, 3 *zhang* 6 *chi*, multiply this by itself to obtain 1,296 *chi*. Multiply by the height, 4 *chi* 5 *cun*, to give 5,832 *chi* and divide by 36 to obtain 162 *chi*. Divide by 1 *chi* 6 *cun* 2 *fen* 分, which is the measure of one *hu*, to get the answer.

[4] Now there is a book on Buddhism with a total of 29 chapters. Each chapter has 63 characters. Find the total number of characters.

Answer: 1,827.

Method: Put down 29 chapters and multiply by 63 characters to get the answer.

[5] Now there is a square chess board with 19 [horizontal and vertical] lines, find the number of chess pieces needed.

Answer: 361.

Method: Put down 19 lines, multiply [the number] by itself to get the answer.

[6] Now there are 98,762 *hu* [of grains to be sent] as land tax. If one cart can transport 50 *hu*, find the number of carts needed.

Answer: 1,975 carts with an odd lot (*ji* 奇) of 12 *hu*.

Method: Put down the taxes, 98,762 *hu*,[3] as the dividend (*shi*) and the cart's capacity of 50 *hu* as the divisor (*fa*). Divide to get the answer.

[7] Now there are 98,766 men. If one soldier is to be conscripted from every 25 men, find the number of soldiers conscripted.
Answer: 3,950 soldiers with an odd lot of 16 men.
Method: Put down 98,766 men as dividend (*shi*) and 25 as divisor (*fa*). Divide to obtain the answer.

[8] Now there are 78,732 *pi* 匹 of thin silk to be divided among 162 persons. Find how much each person gets.
Answer: 486 *pi.*
Method: Put down 78,732 *pi* of thin silk as dividend (*shi*) and 162 persons as divisor (*fa*). Divide to get the answer.

[9] Now there are 36,454 households. If each household sends 2 *jin* 斤 8 *liang* 兩 of silk floss [as payment for tax], find the total amount sent.
Answer: 91,135 *jin.*
Method: Put down 36,454 households. Raise this to tens (*shang shi zhi* 上十之) to obtain 364,540. Multiply by 4 to give 1,458,160 *liang* and divide by 16 to get the answer.

[10] Now there are 91,135 *jin* of silk floss given by 36,454 households. Find the amount from each household.
Answer: 2 *jin* 8 *liang.*
Method: Put down 91,135 *jin* as dividend (*shi*) and 36,454 families as divisor (*fa*). Divide to obtain 2 *jin* and [a remainder]. The remainder of 18,227 *jin* is multiplied by 16 to obtain 291,632 *liang* and divided by the number of households to complete the answer.

[11] Now there are 3,999 *hu* 9 *dou* 6 *sheng* 升 of millet. If every 9 *dou* of millet is exchanged for 1 *hu* of beans, find the amount of beans.

---

[3] Qian [ed. 1963, p. 313, n. 1] says, "The Southern Song version omitted the '2'. There is no mistake in the [*Yongle*] *dadian* edition."

Answer: 4,444 *hu* 4 *dou.*

Method: Put down 3,999 *hu* 9 *dou* 6 *sheng* of millet as dividend (*shi*) and 9 *dou* as divisor (*fa*). Divide to get the answer.

[12] Now there are 2,374 *hu* of millet. If every *hu* is increased by 3 *sheng*, find the total amount of millet.

Answer: 2,445 *hu* 2 *dou* 2 *sheng.*

Method: Put down 2,374 *hu* of millet and multiply by 1 *hu* 3 *sheng* to get the answer.

[13] Now there are 369,980 *hu* 7 *dou* of millet stored in a granary for 9 years. The wastage per year is 3 *sheng* for every *hu*. Find the amount of wastage in 1 year and in 9 years.

Answer: In 1 year, the amount of wastage is 11,099 *hu* 4 *dou* 2 *sheng* 1 *ge* 合. In 9 years, the amount of wastage is 99,894 *hu* 7 *dou* 8 *sheng* 9 *ge.*

Method: Put down 369,980 *hu* 7 *dou* and multiply by 3 *sheng* to obtain the yearly wastage. Next multiply by 9 to get the wastage in 9 years.

[14] Now there is a loan of 57 *jin* of silk to a person. If the yearly interest is 16 *jin*, find the interest per *jin*.

Answer: $4\frac{28}{57}$ *liang.*

Method: Display the interest, 16 *jin* of silk, and multiply by 16 *liang* to obtain 256 *liang*. Divide by the amount of loaned silk, 57 *jin*. The remainder [forming a fraction with the divisor] is reduced to the simplest form to complete the answer.

[15] Now when 3 persons share a cart, there are 2 carts empty. When 2 persons share a cart, 9 persons have to walk. Find the number of persons and the number of carts.

Answer: 15 carts; 39 persons.

Method: Put down 2 carts,[4] multiply by 3 to give 6, add 9, which is the number of persons who have to walk, to obtain 15 carts. To find

---

[4] Qian [ed. 1963 p. 315, n. 1] says, "The various versions give 'put down 2 persons' instead of 'put down 2 carts'. It is now rectified as such in accordance with the given method. The statement 'put down 2 carts' means simply putting down the number of carts."

the number of persons, multiply the number of carts by 2 and add 9, which is the number of persons who have to walk.

[16] Now there are 128,940 *hu* 9 *dou* 3 *ge* of millet. If a man wishes to buy thin silk, 1 *pi* 匹 of which is worth 3 *hu* 5 *dou* 7 *sheng* of millet, find the quantity of thin silk he can get.

Answer: 36,117 *pi* 3 *zhang* 6 *chi*.[5]

Method: Put down 128,940 *hu* 9 *dou* 3 *ge* as dividend (*shi*) and 3 *hu* 5 *dou* 7 *sheng* as divisor (*fa*). Divide to obtain the amount in *pi*. Multiply the remainder by 40 and divide the result by the divisor (*fa*) to complete the answer.

[17] Now there was a woman washing bowls by the river. An officer asked, "Why are there so many bowls?" The woman replied, "There were guests in the house." The officer asked, "How many guests were there?" The woman said, "I don' t know how many guests there were; every 2 persons had [a bowl of] rice, every 3 persons [a bowl of] soup and every 4 persons [a bowl of] meat, 65 bowls were used altogether."

Answer: 60 persons.

Method: Put down 65 bowls, multiply by 12 to obtain 780 and divide by 13 to get the answer.

[18] Now there is a piece of wood whose length is not known. If a rope is used to measure it, the rope has a remainder of 4 *chi* 5 *cun*. If the rope is bent [into two] and is then used to measure it, it is short of 1 *chi*. Find the length of the wood.

Answer: 6 *chi* 5 *cun*.

Method: Put down the remaining part of the rope, 4 *chi* 5 *cun*, and add the deficit, 1 *chi*, to give a total of 5 *chi* 5 *cun*. Double this to obtain 1 *zhang* 1 *chi* and subtract the remainder, 4 *chi* 5 *cun*, to get the answer.

---

[5] This answer is correct. Qian [ed. 1963, p. 315, n. 2] is incorrect in his note, which says, "There is a fractional remainder of $\frac{108}{357}$ *chi* or approximately 3 *cun* after ' 6 *chi*' , but all versions do not take note of the remainder. It is difficult to know whether this is a lacuna."

[19] Now there is an unknown quantity of rice in a container. If the first person takes a half of it, the second person takes a third [of the remainder] and the last person takes a quarter [of what is left], the remaining amount of rice is 1 *dou* 5 *sheng*. Find the original amount of rice.

Answer: 6 *dou*.

Method: Put down the remaining amount of rice, 1 *dou* 5 *sheng*, multiply by 6 to obtain 9 *dou* and divide by 2 to give 4 *dou* 5 *sheng*. Multiply by 4 to obtain 1 *hu* 8 *dou* and divide by 3 to get the answer.

[20] Now there is 1 *jin* of gold which is worth 100,000 *qian* 錢. Find the worth of 1 *liang* [of gold].

Answer: 6,250 *qian*.

Method: Put down 100,000 *qian* and divide by 16 *liang* to get the answer.

[21] Now there is 1 *pi* 疋 of brocade which is worth 18,000 *qian*. Find the worth of 1 *zhang*, 1 *chi*, and 1 *cun* of brocade.

Answer: 1 *zhang* [is worth] 4,500 *qian*, 1 *chi* [is worth] 450 *qian*, and 1 *cun* [is worth] 45 *qian*.

Method: Put down 18,000 *qian*, divide by 4 to obtain the worth of 1 *zhang*. Shift [to the right] (*tui* 退 lit. shift back) by one place and then by another place to obtain the worth of 1 *chi* and 1 *cun* [respectively].

[22] Now there is a piece of land 1,000 *bu* 步 long and 500 *bu* wide. If there is a quail in every [square] *chi* and a small brown speckled bird in every [square] *cun*, find the number of quails and speckled birds.

Answer: 18,000,000 quails, 180,000,000 speckled birds.[6]

Method: Put down the length, 1,000 *bu*, and multiply by the width, 500 *bu*, to obtain 500,000 *bu*. Multiply by 36 to give 18,000,000, which is the number of quails. Raise this to tens (*shang shi zhi* 上 十 之)[7] to obtain the number of speckled birds.

---

[6] The answer for speckled birds is incorrect, it should be 1,800,000,000.

[7] "Raise this to tens (*shang shi zhi*)" should read "raise this to hundreds (*shang bai zhi* 上百之)" to get the correct answer.

[23] Now there is a population of 60,000. The upper [stratum] of the population of 30,000 persons consume 9 *sheng* [each], the middle [stratum] of 20,000 persons consume 7 *sheng* [each], and the lower [stratum] of 10,000 persons consume 5 *sheng* [each]. Find the total amount consumed by the upper, middle and lower [strata] of the population.

Answer: 4,600 *hu*.

Method: Put down the number of persons in each [stratum], multiply each number by the [corresponding] amount consumed [by one person] and add the results to get the answer.

[24] Now there is a square bundle of objects whose outer circumference has 32 pieces. Find the total number of objects.

Answer: 81 pieces.

Method: Put down [32] in two positions. Subtract 8 from the numeral in the upper position and add the remainder to the numeral in the lower position. Continue [in this manner] till there is no remainder, and then add 1 [to the numeral in the lower position] to get the answer.

[25] Now there is a pole whose length is not known. When its shadow is measured, 1 *zhang* 5 *chi* is obtained. When a model staff of length 1 *chi* 5 *cun* is separately erected, its shadow is measured as 5 *cun*. Find the length of the pole.

Answer: 4 *zhang* 5 *chi*.

Method: Put down the length of the shadow of the pole, 1 *zhang* 5 *chi*. Multiply by the length of the model staff, 1 *chi* 5 *cun*, and [the product] is raised to tens (*shang shi zhi*) to obtain 22 *zhang* 5 *chi*. Divide by the shadow of the staff, 5 *cun*, to get the answer.

[26] Now there are an unknown number of things. If we count by threes, there is a remainder 2; if we count by fives, there is a remainder 3; if we count by sevens, there is a remainder 2. Find the number of things.

Answer: 23.

Method: If we count by threes and there is a remainder 2, put down 140. If we count by fives and there is a remainder 3, put down 63. If we count by sevens and there is a remainder 2, put down 30. Add them to obtain 233 and subtract 210 to get the answer. If we count by threes and there is a remainder 1, put down 70. If we count by fives and there is a remainder 1, put down 21. If we count by sevens and there is a remainder 1, put down 15. When [a number] exceeds 106, the result is obtained by subtracting 105.

[27] Now there are six-headed four-legged animals and four-headed two-legged birds [put together]. [A count] above gives 76 heads and [a count] below gives 46 legs. Find the number of animals and birds.
Answer: 8 animals, 7 birds.
Method: Double the number of legs and subtract from this the number of heads. Halve the remainder to get the number of animals. Multiply the number of animals by 4 and subtract [the product] from the number of legs. Halve the remainder to get the number of birds.

[28] Now there are 2 persons, A and B, each of whom holds an unknown amount of money. If A gets one half of B's [money], he has a total of 48. If B gets two thirds (*da ban* 大半) of A's [money], he also has a total of 48. Find how much money each person holds.
Answer: A holds 36, B holds 24.
Method: Use [the method] of *fang cheng* 方程 (rectangular tabulation lit. square procedure) to solve. Put down in the right column 2 for A, 1 for B and 96 for money. Put down in the left column 2 for A, 3 for B and 144 for money. Multiply the left column by the 2 of the right column to obtain 4 for A in the upper position, 6 for B in the middle position, and 288 for money in the lower position.[8] Use the right column to subtract the left column twice so that in the left column, the

---

[8] Qian [ed. 1963, p. 319, n. 2] says, "The printed version of Kong [Jihan] follows Dai Zhen's amendment which adds after [the character] *qian* 錢 the following [instruction] in 23 characters: 'Multiply the right column by the 2 of the left column, obtaining 4 for A in the upper position, 2 for B in the middle position and 192 for money in the lower position.' The original method of the *Sun Zi suanjing* follows the direct subtraction (*zhi chu* 直除) method as given in the *fang cheng* chapter of the *Jiu zhang suanshu*. The statement which follows saying, 'Use the right column to subtract the left column twice', clearly instructs the subtraction of the right column from the left column twice. The additional [instruction] in 23 characters provided by Dai Zhen employing the mutual elimination method by means of cross-multilplication is obviously not Sun Zi's original method."

top position is empty, the middle position has remainder 4 for B, which becomes the divisor (*fa*), and the lower position has remainder 96 for money, which becomes the dividend (*shi*). With the divisor (*fa*) above and the dividend (*shi*) below, [divide] to obtain 24, which is the money B holds. Subtract this from the 96 in the lower position of the right column to give a remainder of 72, which becomes the dividend (*shi*). Let the 2 for A in the upper position of the right column be the divisor (*fa*). With the divisor (*fa*) above and the dividend (*shi*) below, [divide] to obtain 36, which is the money A holds.

[29] Now there are 100 deer [being distributed] in a city. If one household has one deer there is a remainder, and if the remainder is again [being distributed] such that every three households share a deer, then nothing is left. Find the number of households in the city.

Answer: 75 households.

Method: Use [the method of] *ying bu zu* 盈不足 (surplus and deficit). If there are 72 households, there is a surplus of 4 deer. If there are 90 households, there is a deficit of 20 deer. Put down 72 in the upper position of the right column and the surplus, 4, in the lower position of the right column. Put down 90 in the upper position of the left column and the deficit, 20, in the lower position of the left column. Cross-multiply (*wei cheng* 維乘) and add the results to form the dividend (*shi*). Add the surplus and deficit to form the divisor (*fa*). Divide to get the answer.

[30] Now there are 3 fowls which peck a total of 1001 grains of millet. If the chick pecks 1 grain, the hen pecks 2 and the rooster pecks 4, find the [proportional] amounts the owners of the three kinds of fowls have to pay the millet seller.

Answer: The owner of the chick pays 143, the owner of the hen pays 286, the owner of the rooster pays 572.

Method: Put down 1001 grains as the dividend (*shi*). Next take the sum of 7 grains pecked by the 3 fowls as the divisor (*fa*). Divide to obtain 143 grains, which is the amount the chick owner has to pay. Doubling this gives the amount the hen owner has to pay [and doubling again] gives the amount the rooster owner has to pay.

[31] Now there are pheasants and rabbits in the same cage. The top [of the cage] has 35 heads and the bottom has 94 legs. Find the number of pheasants and rabbits.

Answer: 23 pheasants, 12 rabbits.

Method: Put down 35 heads in the upper position and 94 legs in the lower position. Halve the number of legs to obtain 47. Perform repeated [subtractions] by taking away the smaller from the larger [as follows]: the upper 3 is subtracted from the lower 4 and the upper 5 is subtracted from the lower 7; the lower 1 is subtracted from the upper 3 and the lower 2 from the upper 5.[9] The answers are thus obtained.

Another method: Put down the number of heads in the upper position and the number of legs in the lower position. Halve the number of legs and subtract the number of heads, subtract [this result] from the number of heads to get the answers.

[32] Now there is a canal 9 *li* 里 long [with a line of] fish, each 3 *cun* long with its head touching the next fish. Find the number of fish.

Answer: 54,000.

Method: Put down 9 *li*, multiply by 300 *bu* to obtain 2,700 *bu*. Next multiply by 6 *chi* to obtain 16,200 *chi*. Raise this to tens (*shang shi zhi* 上十之) to give 162,000 *cun*. Divide by the length of one fish, 3 *cun*, to get the answer.

[33] Now the distance between Chang'an and Luoyang is 900 *li*. If one revolution of a vehicle's wheel is 1 *zhang* 8 *chi*, find the number of revolutions of the wheel [covering the distance] from Luoyang to Chang'an.

---

[9] Qian [ed. 1963, p. 321, n. 1] says, "In the statement of 'the upper 3 is subtracted from the lower 4 and the upper 5 is subtracted from the lower 7; the lower 1 is subtracted from the upper 3 and the lower 2 from the upper 5', the various versions have inadvertently used the same numbers in the respective subtraction, i.e., '3' instead of '4', '5' instead of '7', '1' instead of '3', and '2' instead of '5'. They are now corrected accordingly. The method covers the subtraction of the upper 35 from the lower 47 to give a remainder 12, and the subtraction of this lower 12 from the upper 35 to yield a remainder 23, which is the number of pheasants."

Answer: 90,000 revolutions.

Method: Put down 900 *li*, multiply by 300 *bu* to obtain 270,000 *bu*. Next multiply by 6 *chi* to obtain 1,620,000 *chi*. Let [the circumference of] the wheel, 1 *zhang* 8 *chi*, be the divisor (*fa*) and divide to get the answer.

[34] Now there is sighted 9 embankments outside; each embankment has 9 trees; each tree has 9 branches; each branch has 9 nests; each nest has 9 birds; each bird has 9 young birds; each young bird has 9 feathers; each feather has 9 colours. Find the quantity of each.

Answer: 81 trees, 729 branches, 6,561 nests, 59,049 birds, 531,441 young birds, 4,782,969 feathers, 43,046,721 colours.

Method: Put down 9 embankments, multiply by 9 to obtain the number of trees. Next multiply by 9 to obtain the number of branches. Next multiply by 9 to obtain the number of nests. Next multiply by 9 to obtain the number of birds. Next multiply by 9 to obtain the number of young birds. Next multiply by 9 to obtain the number of feathers. Next multiply by 9 to obtain the number of colours.

[35] Now there are 3 sisters. The eldest returns once every 5 days, the second returns once every 4 days and the youngest returns once every 3 days. Find the number of days before the 3 sisters meet together.

Answer: 60 days.

Method: Put down on the right 5 days for the eldest sister, 4 days for the second sister and 3 days for the youngest sister. For each numeral, arrange 1 counting rod on the left. [By performing] *wei cheng* 維乘 (cross-multiplication), the number of times each sister returns is obtained. The eldest returns 12 times, the second 15 times and the youngest 20 times. Next multiply each by the [corresponding] number of days to get the answer.

[36] Now there is a pregnant woman whose age is 29. If the gestation period is 9 months, determine the sex of the unborn child.

Answer: Male.

Method: Put down 49, add the gestation period and subtract the age. From the remainder take away 1 representing the heaven, 2 the earth,

3 the man, 4 the four seasons, 5 the five phases, 6 the six pitch-pipes, 7 the seven stars [of the Dipper], 8 the eight winds, and 9 the nine divisions [of China under Yu the Great]. If the remainder is odd, [the sex] is male and if the remainder is even, [the sex] is female.

Appendix 1: Table of Reference for Problems of *Sun Zi Suanjing*

Problems	Page reference in this book	Page reference in Qian [ed. 1963]
Ch. 1 Prob. 1	59, 196	284
Ch. 1 Prob. 2	65–66, 196	285
Ch. 2 Prob. 1	81, 201	295
Ch. 2 Prob. 2	81, 83, 201	295
Ch. 2 Prob. 3	81, 84, 202	295–296
Ch. 2 Prob. 4	81, 85–86, 202	296
Ch. 2 Prob. 5	117, 123, 203	296
Ch. 2 Prob. 6	117, 123, 124, 155, 203	297
Ch. 2 Prob. 7	117, 123, 124, 155, 203	297
Ch. 2 Prob. 8	117, 123, 203	297
Ch. 2 Prob. 9	116, 127, 204	297–298
Ch. 2 Prob. 10	116, 117, 128, 155, 167, 204	298
Ch. 2 Prob. 11	116, 117, 128, 155, 167, 204	298
Ch. 2 Prob. 12	116, 117, 128, 167, 204–205	298–299
Ch. 2 Prob. 13	116, 128, 155, 205	299
Ch. 2 Prob. 14	116, 127, 155, 205	299–300
Ch. 2 Prob. 15	116, 129, 205	300
Ch. 2 Prob. 16	116, 126, 206	300
Ch. 2 Prob. 17	116, 128, 206	300
Ch. 2 Prob. 18	116, 128, 206	301
Ch. 2 Prob. 19	30, 93–94, 116, 121, 126, 206–207	301–302
Ch. 2 Prob. 20	30, 93, 103, 116, 121, 128, 207–208	302
Ch. 2 Prob. 21	116, 128, 208	302–303
Ch. 2 Prob. 22	116, 129, 208–209	303
Ch. 2 Prob. 23	116, 129, 155, 209	303–304
Ch. 2 Prob. 24	129–130, 209–210	304–305
Ch. 2 Prob. 25	131, 156–157, 165, 210	305
Ch. 2 Prob. 26	132, 210–211	305–306
Ch. 2 Prob. 27	116, 154, 155, 211	306–307
Ch. 2 Prob. 28	116, 143–144, 168, 211	307
Ch. 3 Prob. 1	118, 134–135, 165, 166, 213–214	309
Ch. 3 Prob. 2	118, 164, 214	310–311
Ch. 3 Prob. 3	116, 117, 129, 167, 214	311
Ch. 3 Prob. 4	27, 155, 164, 214	311
Ch. 3 Prob. 5	27, 127, 165, 214	311
Ch. 3 Prob. 6	118, 155, 165, 166, 214–215	311–312

Appendix 1 (*continued*)

Problems	Page reference in this book	Page reference in Qian [ed. 1963]
Ch. 3 Prob. 7	155, 164, 215	312
Ch. 3 Prob. 8	116, 136, 215	312
Ch. 3 Prob. 9	27, 28, 117, 136, 165, 166, 215	312
Ch. 3 Prob. 10	117, 136, 165, 166, 215	312–313
Ch. 3 Prob. 11	117, 125, 168, 215	313
Ch. 3 Prob. 12	118, 155, 216	313
Ch. 3 Prob. 13	118, 155, 167, 216	313–314
Ch. 3 Prob. 14	30, 117, 155, 168, 216	314
Ch. 3 Prob. 15	137, 216	314
Ch. 3 Prob. 16	116, 117, 125, 168, 217	314–315
Ch. 3 Prob. 17	28, 137, 217	315
Ch. 3 Prob. 18	116, 126, 217	315
Ch. 3 Prob. 19	118, 138, 218	315–316
Ch. 3 Prob. 20	117, 125, 168, 218	316
Ch. 3 Prob. 21	116, 125, 168, 218	316
Ch. 3 Prob. 22	116, 127, 136, 218	316–317
Ch. 3 Prob. 23	118, 136, 219	317
Ch. 3 Prob. 24	155–156, 219	317
Ch. 3 Prob. 25	116, 136, 156–157, 219	317
Ch. 3 Prob. 26	26, 138–140, 219–220	318
Ch. 3 Prob. 27	157, 220	318
Ch. 3 Prob. 28	146, 220–221	318–319
Ch. 3 Prob. 29	140–141, 221	319
Ch. 3 Prob. 30	138, 221	319–320
Ch. 3 Prob. 31	157, 222	320
Ch. 3 Prob. 32	116, 126, 136, 222	320–321
Ch. 3 Prob. 33	27, 116, 128, 163, 222–223	321
Ch. 3 Prob. 34	158, 223	321–322
Ch. 3 Prob. 35	158–159, 223	322
Ch. 3 Prob. 36	27, 159–160, 223–224	322

Appendix 2: Chronology of Dynasties

Xia dynasty		About 21st century to 16th century BC
Shang dynasty		About 16th century to 11th century BC
Zhou dynasty	Western Zhou	11th century to 771 BC
	Eastern Zhou	770–221 BC
	Spring and Autumn period	770–476 BC
	Warring States period	475–221 BC
Qin dynasty		221–206 BC
Han dynasty	Western Han	206 BC–24 AD
	Eastern Han	25–220 AD
Three Kingdoms	Wei	220–265 AD
	Shu	221–263 AD
	Wu	222–280 AD
Jin dynasty	Western Jin	265–316 AD
	Eastern Jin	317–420 AD
Northern and Southern dynasties	Southern dynasties:	
	Song	420–479 AD
	Qi	479–502 AD
	Liang	502–557 AD
	Chen	557–589 AD
	Northern dynasties:	
	Northern Wei	386–534 AD
	Eastern Wei	534–550 AD
	Northern Qi	550–577 AD
	Western Wei	535–556 AD
	Northern Zhou	557–581 AD
Sui dynasty		581–618 AD
Tang dynasty		618–907 AD
Five dynasties	Later Liang	907–923 AD
	Later Tang	923–936 AD
	Later Jin	936–946 AD
	Later Han	947–950 AD
	Later Zhou	951–960 AD
Song dynasty	Northern Song	960–1126 AD
	Southern Song	1127–1279 AD
	Jin	1115–1234 AD
Yuan dynasty		1271–1368 AD
Ming dynasty		1368–1644 AD
Qing dynasty		1644–1911 AD
Republic		1912–1949 AD

Bibliography

Ang Tian Se 1969. A study of the mathematical manual of Chang Ch' iu-chien. Unpublished M.A. dissertation, University of Malaya.

Ang Tian Se 1977. Chinese computation with counting rods. *Papers on Chinese Studies*, University of Malaya **1**, 97–109.

Ang Tian Se 1978. Chinese interest in right-angled triangles. *Historia Mathematica* **5**, 253–266.

Ang Tian Se 1979. I-Hsing (683–727 AD): His life and scientific work. Unpublished Ph.D dissertation, University of Malaya.

Bacon, Francis 1905. *Philosophical works*. Edited by Ellis & Spedding. London: Routledge.

Bai Shangshu 白尚恕 1983. *Jiu zhang suanshu zhushi* 九章算術注釋 (Annotations on *Jiu zhang suanshu*). Beijing: Kexue chubanshe.

Bell, E.T. 1940. *The development of mathematics*. 2nd ed. 1945. New York: McGraw-Hill.

Berezkina, E.I. 1957. Drevnekitajskij trakat "Matematika v devjati knigach" (The ancient Chinese mathematical work "Nine chapters on the mathematical art"). *Istoriko-Matematiceskie Issledovanija* **10**, 423–584.

Boncompagni, Baldassarre 1857. *Liber Abbaci di Leonardo Pisano*. Rome: Tipografia delle Scienze Matematiche e Fisiche.

Boyer, Carl B. 1944. Fundamental steps in the development of numeration. *Isis* **35**, 153–168.

Cajori, Florian 1893. *A history of mathematics*. 3rd ed. 1980. New York: Chelsea Publishing Co.

Cajori, Florian 1896. *A history of elementary mathematics*. 6th ed. 1953. London: Macmillan.

Cajori, Florian 1928. *A history of mathematical notations*. Vol. 1: *Notations in elementary mathematics*. London: Open Court.

Chen Zunguei 陳遵嬀 1955. *Zhongguo gudai tianwenxue jian shi* 中國古代天文學簡史 (A short history of ancient Chinese astronomy). Shanghai: Shangwu yinshuguan.

Ch' en, Kenneth 1964. *Buddhism in China. A historical survey*. Princeton: Princeton University Press.

Cheng Chin-te 1925. The use of computing rods in China. *The American Mathematical Monthly* **32**, 492–499.

Cheng Te-k' un 1983. *Studies in Chinese art*. Hong Kong: The Chinese University Press.

Cohn, P.M. 1958. *Linear equations*. London: Routledge & Kegan Paul.

Colebroke, H.T. 1817. *Algebra, with arithmetic and mensuration. From the Sanscrit of Brahmegupta and Bhascara*. London: John Murray.

Crossley, John N. & Henry, Alan S. 1990. Thus spake al-Khwarizimi: A translation of the text of Cambridge University Library Ms. Ii.vi.5. *Historia Mathematica* **17**, 103–131.

Dantzig, Tobias 1930. *Number. The language of science*. 3rd ed. 1947. London: George Allen & Unwin.

Datta, B. & Singh, A.N. 1935. *History of Hindu mathematics*. Part 1: *Numeral notation and arithmetic*. Single vol. ed. 1962. Bombay: Asia Publishing House.

de Hungaria, Georgij 1499. *Arithmeticae summa tripartita*. Facsimile with introduction by A.J.E.M. Smeur 1965. Nieuwkoop (Holland): B. de Graaf.

Dickson, Leonard Eugene 1957. *Introduction to theory of numbers*. New York: Dover Publications.

Du Shiran 杜石然 1991. Lüe lun zhongguo gudai shuxue shizhong de weizhizhi sixiang 略論中國古代數學史中的位值制思想 (A brief discussion on the concept of place value system in the history of ancient Chinese mathematics). In *Zhongguo gudai kexue shilun xu pian* 中國古代科學史論續篇 pp. 157–171. Kyoto: Jingdu daxue renwen kexue yanjiusuo yanjiu baogao.

Flegg, Graham (ed.) 1989. *Numbers through the ages*. London: Macmillan.

Gillon, B.S. 1977. Introduction, translation, and discussion of Chao Chün-Ch' ing' s "Notes to the diagrams of short legs and long legs and of squares and circles". *Historia Mathematica* **4**, 253–293.

Gow, J. 1923. *A short history of Greek mathematics*. New York: Stechert.

Gupta, R.C. 1989. On some rules from Jaina mathematics. *Ganita-Bharati* **11**, 18–26.

Heath, T.L. 1921. *A history of Greek mathematics*. 2 vols. Oxford: Clarendon Press.

Hill, G.F. 1915. *The development of Arabic numerals in Europe exhibited in sixty-four tables*. Oxford: Clarendon Press.

Hodder, James 1672. *Arithmetick or the necessary art made most easy*. 10th ed. London.

Hoe, J. 1977. *Les systèmes d' équations polynômes dans le Siyuan yujian (1303)*. France: Collège de France, Institut des Hautes Études Chinoises.

Karpinski, L.C. 1925. *The history of arithmetic*. Chicago & New York: Rand McNally.

Kaye, G.R. 1907. Notes on Indian mathematics. Arithmetical notation. *Journal of the Asiatic Society of Bengal* **3**, 475–508.

Lam Lay Yong 1969a. On the existing fragments of Yang Hui' s Hsiang Chieh Suan Fa. *Archive for History of Exact Sciences* **6**, 82–88.

Lam Lay Yong 1969b. The geometrical basis of the ancient Chinese square-root method. *Isis* **61**, 92–102.

Lam Lay Yong 1974. Yang Hui' s commentary on the *ying nu* chapter of the *Chiu Chang Suan Shu*. *Historia Mathematica* **1**, 47–64.

Lam Lay Yong 1977. *A critical study of the Yang Hui suan fa. A thirteenth-century Chinese mathematical treatise*. Singapore: Singapore University Press.

Lam Lay Yong 1979. Chu Shih-chieh' s Suan-hsüeh ch' i-meng (Introduction to mathematical studies). *Archive for History of Exact Sciences* **21**, 1–31.

Lam Lay Yong 1980. The Chinese connection between the Pascal Triangle and the solution of numerical equations of any degree. *Historia Mathematica* **7**, 407–424.

Lam Lay Yong 1982. Chinese polynomial equations in the thirteenth century. In Li Guohao et al. (eds.) *Explorations in the history of science and technology in China* (pp. 231–272). Shanghai: Shanghai Chinese Classics Publishing House.

Lam Lay Yong 1986. The conceptual origins of our numeral system and the symbolic form of algebra. *Archive for History of Exact Sciences* **36**, 183–195.

Lam Lay Yong 1986a. The development of polynomial equations in traditional China. *Mathematical Medley* **14**, no. 1, 9–34, Singapore Mathematical Society.

Lam Lay Yong 1987. Linkages: Exploring the similarities between the Chinese rod numeral system and our numeral system. *Archive for History of Exact Sciences* **37**, 365–392.

Lam Lay Yong 1987a. Geometical algebra in quadratic equations in China and the counting rod system. *Bulletin of Chinese Studies* **1**, no. 2, 153–170, University of Hong Kong.

Lam Lay Yong 1988. A Chinese genesis: Rewriting the history of our numeral system. *Archive for History of Exact Sciences* **38**, 101–108.

Lam Lay Yong 1994. Jiu Zhang Suanshu (Nine Chapters on the Mathematical Art): An Overview. *Archive for History of Exact Sciences*, **47**, 1–51.

Lam Lay Yong & Ang Tian Se 1986. Circle measurements in ancient China. *Historia Mathematica* **13**, 325–340.

Lam Lay Yong & Ang Tian Se 1987. The earliest negative numbers: How they emerged from a solution of simultaneous linear equations. *Archives Internationales D' Histoire des Sciences* **37**, 222–262.

Lam Lay Yong & Shen Kangshen 1984. Right-angled triangles in ancient China. *Archive for History of Exact Sciences* **30**, 87–112.

Lam Lay Yong & Shen Kangshen 1989. Methods of solving linear equations in traditional China. *Historia Mathematica* **16**, 107–122.

Levey, Martin & Petruck, Marvin 1965. *Kūshyār ibn Labbān. Principles of Hindu reckoning.* Madison & Milwaukee: University of Wisconsin.

Li Di 李迪 1984. *Zhongguo shuxueshi jianbian* 中國數學史簡編 (A concise history of Chinese mathematics). Shenyang: Liaoning renming chubanshe.

Li Wenlin & Yuan Xiangdong 1983. The Chinese remainder theorem. In *Ancient China's technology and science* (pp. 99–110). Compiled by the Institute of the History of Natural Sciences, Chinese Academy of Sciences. Beijing: Foreign Languages Press.

Li Yan 李嚴 1954. *Zhongguo gudai shuxue shiliao* 中國古代數學史料 (A history of ancient Chinese mathematics). Shanghai: Kexue jishu chubanshe.

Li Yan 李嚴 & Du Shiran 杜石然 1963. *Zhongguo gudai shuxue jianshi* 中國古代數學簡史 (A concise history of ancient mathematics in China). 2 vols. Beijing: Zhonghua shuju.

Li Yan & Du Shiran 1987. *Chinese mathematics. A concise history.* Translated by J.N. Crossley & A.W.C. Lun. Oxford: Clarendon Press.

Libbrecht, Ulrich 1973. *Chinese mathematics in the thirteenth century. The Shu-shu chiu-chang of Ch' in Chiu-shao.* Massachusetts: MIT Press.

Libbrecht, Ulrich 1982. Mathematical manuscripts from the Tunhuang Caves. In Li Guohao et al. (eds.) *Explorations in the history of science and technology in China* (pp. 203–229). Shanghai: Shanghai Chinese Classics Publishing House.

Martzloff, Jean-Claude 1988. *Histoire des mathématiques Chinoises.* Paris: Masson.

Mei Rongzhao 1983. The decimal place-value numeration and the rod and bead arithmetics. In *Ancient China's technology and science* (pp. 57–65). Compiled by the Institute of the History of Natural Sciences, Chinese Academy of Sciences. Beijing: Foreign Languages Press.

Menninger, Karl 1969. *Number words and number symbols. A cultural history of numbers.* Translated by Paul Broneer from the revised 1958 German edition. Massachusetts: MIT Press.

Mikami, Y. 1914. On the Japanese theory of determinants. *Isis* **2**, 9–36.

Needham, Joseph 1959. *Science and civilisation in China.* Vol. 3: *Mathematics and the sciences of the heavens and the earth.* Cambridge: Cambridge University Press.

Qian Baocong 錢寶琮 1927. Jiu zhang suanshu ying bu zu shu liu zhuan Ouzhou kao 九章算術盈不足術流傳歐洲考 (On the transmission of the rule of 'surplus and deficit" in *Jiu zhang suanshu* to Europe). *Kexue* 科學 **12**, no. 6, 707.

Qian Baocong (ed.) 1963. *Suanjing shi shu* 算經十書 (Ten mathematical classics). Beijing: Zhonghua shuju.

Qian Baocong 1964. *Zhongguo shuxueshi* 中國數學史 (A history of Chinese mathematics). Beijing: Kexue chubanshe. 2nd edition 1981.

Qian Baocong et al. 1966. *Song Yuan shuxueshi lunwenji* 宋元數學史論文集 (Collected essays on the history of mathematics of the Song-Yuan period). Beijing: Kexue chubanshe.

Saidan, A.S. 1978. *The Arithmetic of Al-Uqlīdisī.* Dordrecht: D. Reidel.

Shen Kangshen 沈康身 1986. *Zhong suan daolun* 中算導論 (Instructions on Chinese mathematics). Shanghai: Shanghai jiaoyu chubanshe.

Smeur, A.J.E.M. 1978. The Rule of False applied to the quadratic equation, in three sixteenth century arithmetics. *Archives Internationales d' Histoire des Sciences* **28**, 66–101.

Smith, D.E. 1923. *History of mathematics.* Vol. 1: *General survey of the history of elementary mathematics.* Boston: Ginn & Co.

Smith, D.E. 1925. *History of mathematics.* Vol. 2: *Special topics of elementary mathematics.* Boston: Ginn & Co.

Smith, D.E. & Ginsburg, J. 1937. *Numbers and numerals.* New York: Teachers College, Columbia University.

Smith, D.E. & Karpinski, L.C. 1911. *The Hindu-Arabic numerals.* Boston: Ginn & Co.

Srinivasiengar, C.N. 1967. *The history of ancient Indian mathematics.* Calcutta: World Press.

Swetz, F.J. 1987. *Capitalism and arithmetic. The new math of the 15th century including the full text of the Treviso Arithmetic of 1478.* Illinois: Open Court.

Swetz, F.J. & Kao, T.I. 1977. *Was Pythagoras Chinese? An examination of right triangle theory in ancient China.* Pennsylvania: Pennsylvania State University Press.

Vogel, Kurt 1963. *Mohammed ibn Musa Alchwarizi's algorismus.* Aalen: Otto Zeller Verlagsbuchhandlung.

Vogel, Kurt 1968. *Neun bucher arithmetischer technik.* Braunschweig: Friedr. Vieweg & Sohn.

Wang Guowei 1928. Chinese foot-measures of the past nineteen centuries. Translated by A.W. Hummel & Fung Yu-lan. *Journal of the North-China Branch of the Royal Asiatic Society* **59**, 111–123.

Wang Ling 1956. The Chiu chang suan shu and the history of Chinese mathematics during the Han dynasty. Unpublished Ph.D. dissertation, University of Cambridge.

Wang Ling 1964. The date of the Sun Tzu suan ching and the Chinese remainder problem. *Proceedings of 10th International Congress of the History of Science (Ithaca, 1962)* **1**, 489–492.

Wang Ling & Needham, Joseph 1955. Horner's method in Chinese mathematics: Its origins in the root-extraction procedures of the Han dynasty. *T'oung Pao* **43**, 345–401.

Wilder, Raymond L. 1968. *Evolution of mathematical concepts. An elementary study.* New York: John Wiley.

Wright, Arthur F. 1959. *Buddhism in Chinese history.* Stanford: Stanford University Press.

Wu Chengluo 吳承洛 1937. *Zhongguo du liang heng shi* 中國度量衡史 (A history of Chinese metrology). Shanghai: Shangwu yinshuju. 2nd edition 1957.

Wu Wenjun 吳文俊 (ed.) 1982. *Jiu zhang suanshu yu Liu Hui* 九章算術與劉徽 (The Jiu zhang suanshu and Liu Hui). Beijing: Beijing shifan daxue chubanshe.

Xu Chunfang 許蒓舫 1965. *Zhongguo suanshu gushi* 中國算術故事 (Stories of Chinese arithmetic). Beijing: Zhongguo qinglian chubanshe.

Xu Xintong 許鑫銅 1987. Sun Zi suanjing shou chuang kaifangfa zhong de "chao wei tui wei ding wei fa" 孫子算經首創開方法中的 "超位退位定位法" (Master Sun's Arithmetical Manual is the first to originate "the leap forward and regress method of place determination" in extraction.) *Journal of East China Normal.* Natural Science Edition no. 1, 22–27.

Yan Dunjie 嚴敦傑 1937. Sun Zi suanjing yanjiu 孫子算經研究 (A study of *Sun Zi suanjing*). *Xue yi zazhi* 學藝雜誌 **16**, no. 3, 311–327.

Yang Lien-sheng 1961. *Studies in Chinese institutional history.* Massachusetts: Harvard-Yenching Institute Studies XX. 3rd printing 1969.

Zhang Yinlin 張蔭麟 1927. Jiu zhang ji liang Han zhi shuxue 九章及兩漢之數學 (The *Jiu zhang* and mathematics in the two Han dynasties). *Yanjing xuebao* 燕京學報 **2**, 301.

Supplementary Bibliography for Books in Chinese

Chouren zhuan 疇人傳 (Biographies of mathematicians and astronomers). Compiled by Ruan Yuan 阮元 1799. Shanghai: Shangwu yinshuguan, 1955.

Daode jing 道德經 (Canon of the *dao* and its virtues). Attributed to Lao Zi 300 BC. In *Si bu congkan* 四部叢刊. Shanghai: Shangwu yinshuguan, 1919.

Jin shu 晉書 (Standard history of the Jin dynasty). Compiled by Fang Xuanling 房玄齡 et al. 635 AD. Beijing: Zhonghua shuju, 1975.

Jiu Tang shu 舊唐書 (Old standard history of the Tang dynasty). Compiled by Liu Xu 劉昫 945 AD. Beijing: Zhonghua shuju, 1975.

Qian Han shu 前漢書 (Standard history of the Western Han dynasty). Compiled by Ban Gu 班固 and after his death in 92 AD by his sister Ban Zhao 班昭 c.100 AD. Beijing: Zhonghua shuju, 1975.

Sui shu 隋書 (Standard history of the Sui dynasty). Compiled by Wei Zheng 魏徵 et al. 656 AD. Beijing: Zhonghua shuju, 1973.

Taiping yu lan 太平御覽 (Imperial encyclopedia of the Taiping reign period). Edited by Li Fang 李昉 983 AD. In Guo xue jiben congshu 國學基本叢書. Taipei: Shangwu yinshuguan, 1968.

Wei shu 魏書 (Standard history of the Wei dynasty). Compiled by Wei Shou 魏收 572 AD. Beijing: Zhonghua shuju, 1974.

Xin Tang shu 新唐書 (New standard history of the Tang dynasty). Compiled by Ouyang Xiu 歐陽修 & Song Qi 宋祁 1061. Beijing: Zhonghua shuju, 1975.

Yongle dadian 永樂大典 (Great encyclopedia of the Yongle reign period). Edited by Xie Jin 解縉 1407. Beijing: Zhonghua shuju, 1960.

Index